UNLOCKING THE POWER OF OPC-UA

A Guide to Transforming Industries

Praveen Kumar Singh

Smitha Rao

Chatrapathi G V

Andreas Faath

INDIA • SINGAPORE • MALAYSIA

ISBN 979-8-89186-541-9

Legal Disclaimer

2023 All Rights Reserved @ ISBN 979-8-89186-541-9

UNLOCKING THE POWER OF OPC UA

Abstract

Explore the transformative potential of OPC UA in this authoritative guide. Tailored for industry leaders navigating a dynamic digital terrain, this book offers a strategic roadmap to enhance operational efficiency and resilience. Debunking prevalent misconceptions about OPC UA's relevance to industries of all scales, it sheds light on how leading enterprises have leveraged this technology for enduring advantages.

Don't just adapt—lead.

Discover how embracing OPC UA can place your organization at the forefront of innovation.

Become future-ready now!

Table of Contents

Terms of Use

By accessing, purchasing, or using this book, you agree to be bound by the following terms and conditions. Please read them carefully.

1. **Intellectual Property Rights**

   - The content within this book, including but not limited to written text, graphics, images, and illustrations, is the collective intellectual creation of its authors. All the content and the ideas are derived from their practical experience and exposure in their respective fields. This intellectual property belongs solely to the authors and Utthunga Technologies Pvt. Ltd. (hereinafter referred to as "Utthunga").

2. **Use of External References and Attribution**

   - This publication may include references sourced from various external resources available online. All original credit and associated intellectual property rights for these references remain with the rightful owners. Neither Utthunga nor the authors claim ownership of these external references.

3. **Content Discretion**

   - The ideas, suggestions, and recommendations expressed in this book are subject to personal interpretation. Any disagreements or differences in opinion regarding the content of this book are the sole responsibility of the reader. Utthunga and the authors do not assume liability for any interpretations or actions taken by the reader based on the content of this book.

4. **Acknowledgment of Legal Ownership and Attribution**

- This book may include direct or indirect references, such as content, illustrations, diagrams, and other supporting materials, to enhance and support the original content. Utthunga fully respects and acknowledges all legal ownership and intellectual property rights of such references to their rightful owners.

By using this book, you acknowledge that you have read, understood, and agreed to these Terms of Use.

...The Team

The creation of this book has been a collective endeavor, a testament to the spirit of collaboration and shared vision. We are profoundly grateful to a remarkable group of individuals whose expertise, dedication, and passion have been instrumental in transforming a complex array of ideas into a comprehensive and accessible resource on industrial technology. From the depths of detailed research to the finesse of editorial precision and from the nuances of design to the rigors of administrative coordination, each contribution has been a vital thread in the tapestry of this work.

Our special thanks to Mr. Dinesh Thukaram, whose years of industrial experience have infused this book with invaluable real-world perspectives. His insights and practical wisdom have greatly enriched the content, making it not only informative but also relevant to the readers.

We extend our sincere appreciation to Supriya Jain, whose exceptional skill in articulating and refining our ideas played a pivotal role in shaping the narrative of this book. Her ability to distill complex discussions and concepts into clear, compelling content has been nothing short of remarkable. Supriya's contributions have truly been the crucible in which our raw ideas were transformed into the polished insights that grace these pages.

This book is not merely a reflection of knowledge, but a mosaic crafted by the hands and hearts of many. We extend our heartfelt thanks to all those who have journeyed with us in making this publication a reality.

Contributors

Nithin S P

Supriya Jain

Suraj S

Ishac Chemmala

Gopal B Dambal

Christopher Liehr

Jakob Albert

Maheshwari B

Vinay BR

Vijay Kadkol

Soujanya Viswa

Sadatulla Zishan

Foreword

Krishnan K M
Founder & Managing Director
Utthunga Technologies Pvt. Ltd.

In the rapidly evolving industrial automation sector, this publication on OPC UA Technologies emerges as a critical guide. It focuses on OPC UA's role in overcoming industry challenges like vendor lock-in and improving interoperability, which is essential for optimizing resources and enhancing operational efficiency.

This book is based on Utthunga's journey in creating globally recognized OPC-based products. It leverages our extensive experience to offer a deep dive into OPC technologies' applications across industries, clarifying their adaptability and ease of integration into existing systems.

Drawing from Utthunga's problem-solving experiences, the book's practical approach showcases real-life OPC technology applications in improving operations. It's crafted for a broad audience within the industrial ecosystem, aiming to demystify OPC technologies for both technical and decision-making readers.

The book acknowledges OPC's role within a broader industrial context and provides a nuanced view of its contribution to unified interoperability. This perspective is crucial for companies looking to stay competitive in a swiftly changing industry landscape.

This work represents a collective effort by leading OPC experts, combining their in-depth knowledge to deliver a comprehensive guide to navigating modern industrial automation complexities. As we approach Industry 5.0, this book is a pivotal resource, encouraging strategic advancement in industrial automation.

We hope this guide broadens your understanding of OPC UA and equips you to leverage these technologies for your organization's future success.

Happy Reading!
Krishnan K M
www.utthunga.com

...The Beginning!

Limited edition Pre-release of the book at SPS 2023, Nuremberg - Germany

Acknowledgement

Stefan Hoppe
President & Executive Director - OPC Foundation

I'm excited to present this crucial guide on how OPC UA is revolutionizing industries worldwide as part of the IIoT and Industry 4.0 evolution. By centralizing itself in industry transformation, OPC UA streamlines operations and boosts value across the operational spectrum, marking a significant shift towards unified communication and secure infrastructure.

OPC UA plays a critical role in enabling efficient, secure data exchange between devices and systems, which is vital to the performance of modern, advanced industrial systems. It supports integrated, real-time decisions, paving the way for decentralized operations.

This publication thoroughly and accessibly details OPC UA's transformative impact, offering organizations a clear path to revolutionize their operations for enhanced efficiency, resilience, and preparedness for the future. Emphasizing OPC UA as the industrial automation standard sets the stage for future innovations.

I've seen first-hand how Utthunga's innovative strategies have driven success across industries, a testament to the authors' combined expertise. With rich case studies drawn from their extensive experience, this book addresses the global market's complexities, providing practical advice to excel in today's competitive environment.

Congratulations to the authors and contributors for integrating their extensive knowledge and insights into this book. It equips readers with a deep understanding of OPC UA's capabilities and guides businesses toward achieving global standards of excellence.

Stefan Hoppe
www.opcfoundation.org

Why Read This Book?

In an age where industrial digitization is not just an advantage but a necessity, this book on OPC UA is an indispensable guide for field experts and decision-makers. This comprehensive tome demystifies the OPC UA technology, a cornerstone in the edifice of Industry 4.0, by presenting it not merely as a protocol but as a strategic technology framework.

With its in-depth analysis, practical scenarios, and real-world use cases, this book cuts through the noise, debunking common myths and misconceptions surrounding OPC UA. By doing so, it equips decision-makers with the knowledge to make informed choices, ensuring their businesses are not just keeping pace but setting the trend in an increasingly competitive industrial landscape.

The true strength of "Mastering OPC UA" lies in its holistic approach focusing on Industry Use Cases. Each chapter is structured to provide not only theoretical insight but also actionable information that can be directly applied to a variety of industrial contexts. Whether you operate a small industrial unit or helm a multinational conglomerate, the scalability and flexibility of OPC UA, as presented in this book, are bound to resonate.

The case studies, drawn from many sectors, offer a panoramic view of the potential and versatility of OPC UA technologies, showcasing how they can be harnessed to streamline operations, enhance efficiency, and, ultimately, drive business growth.

In essence, this book is more than just a collection of guidelines; it's a roadmap to the industry's future -one that promises to guide companies of all sizes through the intricate maze of digital transformation.

Visible Challenges
Real Challenges

The Industrial Automation Challenge

It's a busy day on the shop floor of a bottling plant, and the assembly lines are buzzing. Robots are busy loading pallets, and AGVs are moving purposefully. A new machine has been installed to check the quality of bottles and discard rejects. It looks like a picture-perfect smart factory. That is, until the management wants a view of the quality metrics.

Frank Edger, the field engineer, needs to submit the quality report. But the new machine has its own configuration and process monitoring software tool. The data from this machine isn't being fed into a central reporting tool. There is no easy way for Frank to know the total number of inspected items per batch, rejected item count, etc. To generate a report, he must manually input the data into an existing system – an exercise that takes much of his productive time. In addition, to make the report more comprehensive, he needs data from a production machine that provides logs in a proprietary format. He must convert these two logs to Excel before he can merge them and calculate the required parameters. What a waste!

This is not an isolated case. Facilities across industries - be it for discrete or process manufacturing automation - are plagued with these inefficiencies because their systems don't talk to each other.

> **Challenge: Secure and scalable communication challenge in remote sites:**
>
> Connecting remote field devices to enterprise applications to create a single data source despite different hardware variants of edge gateways and different server modes.
>
> **Challenge: Effective data aggregation challenge for Process Control Systems:** Connecting the two individual controllers (Source 'A' and Source 'B') to aggregate data from several devices, PLCs, and servers without compromising on the redundancy and performance compliance requirements.

Every industry stakeholder is asking two key questions:

- How can we effortlessly move data between enterprise systems, monitoring devices, and sensors?
- How can we transform data into wisdom and deliver it to both humans and machines for intelligent operations?

• • •

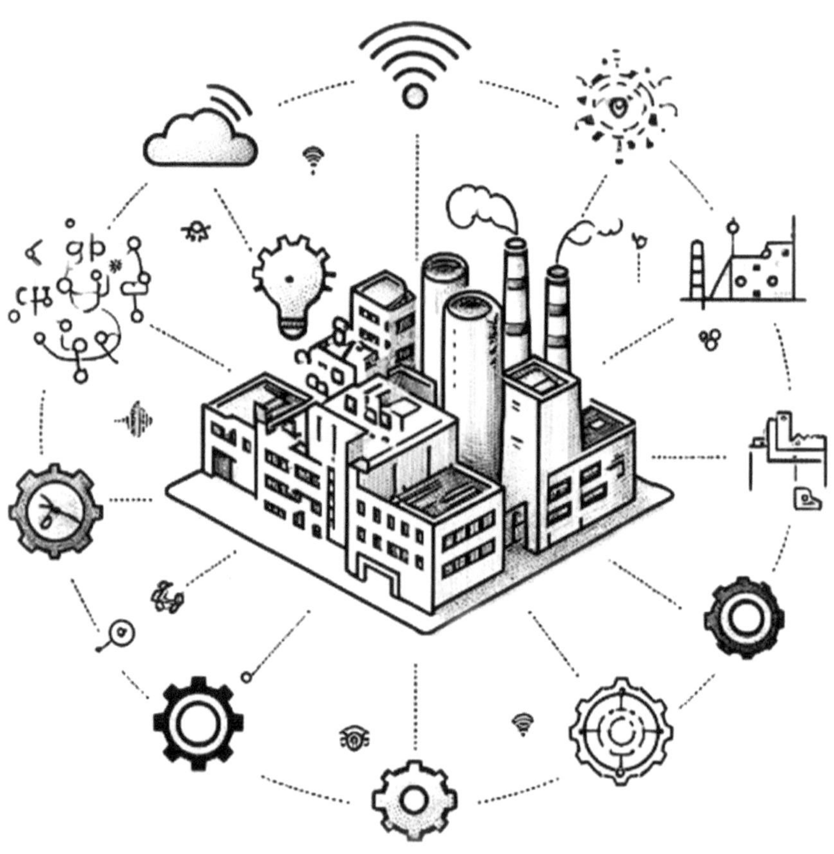

Interoperability

THE HOLY GRAIL OF INDUSTRIAL AUTOMATION

In 2017, a headline by The Economist[1] proclaimed that "The world's most valuable resource is no longer oil, but data." That is true in the industrial context too. Data has the power to write the next chapter in the industrial transformation story. One where connected devices, systems, and people communicate with contextual, real-time data not just in one facility but across the organization's entire value chain.

Imagine factories and facilities where there is no unexpected downtime. Where predictive maintenance ensures the upkeep of every asset in real-time. Where physical and digital systems work together to amplify productivity, efficiency, and flexibility of operations. Where everything is tracked, accounted for, and reported. Where people are safe from hazardous conditions, and waste is minimized.

Today with smart technologies, this vision of connected industrial operations is possible. In fact, industrial automation is sweeping across enterprises, connecting machines, systems, and people. Manufacturers and other industrial players are prioritizing asset digitalization to be more responsive to evolving market needs. The next level of this evolution is digital transformation.

However, the quest for digital transformation requires a new way of thinking about process organization and changes along the entire value chain. Ultimately, companies prioritize value creation, and the question at the forefront is cost-effectiveness and interoperability. While digital ecosystems serve as the foundation for digital transformation, intelligent and connected production requires more than just online connectivity. And the manifestations of digital transformation vary from one company to another, sometimes significantly. There is no one-size-fits-all blueprint for digitalization.

Yet, the potential of Industry 4.0 to hone competitive advantage, deliver better experiences, and create net-new business models is curtailed by data siloes. The inability to exchange data - across hardware, software, and services - securely and independently of platforms and protocols remains a significant challenge for the industry. Many of the systems use different field communication protocols and require expensive custom code to interface with other systems.

Interoperability Issues		
Data transfer between systems	Compatibility issues of different software versions	Lack of common language or syntax
Non-standardized documentation	Inadequate application testing	Inconsistent data formats or standards
31% of organizations lack confidence in interoperability of connected devices and related technologies[2]		**90%** of organizations that use operational technology (OT) systems have experienced some sort of cyber incident in the past year.[3]

Without this data exchange, communication and collaboration suffer, creating operational inefficiencies. Today, this data exchange happens via complex software frameworks that, while being extremely costly, also create a vendor lock-in challenge for enterprises.

To create industrial utopia, we need to securely connect devices and systems to allow an open flow of information between them. Data flowing seamlessly between machines is critical to realize the vision of Industry 4.0 - real-time optimization of the production process and personalization. But even investments in IIOT technologies are no guarantee of interoperability.

So how can we enable a seamless exchanging of information with syntax and semantics understandable by all the heterogeneous systems involved without ripping and replacing existing investments? The answer lies in Open Platform Communications Unified Architecture or OPC UA.

TAKING CONNECTIVITY TO THE NEXT LEVEL WITH OPC UA

The OPC Foundation developed OPC UA as the next-generation machine-to-machine communication protocol that facilitates multi-vendor, multi-platform, secure, and reliable interoperability. Microsoft has hailed this specification as "the de-facto communication technology for Industry 4.0, essential to reaching this next level of connectivity in manufacturing facilities.

As a standard interface for device and system integration, providing end-to-end encryption for security, cloud integration for scalability, and a smooth migration path for legacy systems, OPC UA has become a necessity for modern industrial operations.

OPC UA
• Provides a standard interface that allows seamless integration of different devices and systems, regardless of the industry or underlying technology
• Ensures data security, offering end-to-end encryption, authentication, and access control mechanisms
• Can be supported on a wide range of components, from low-end sensors to high-end servers
• Allows easy modeling of objects, making it useful for a variety of applications without the need for added IT infrastructure
• Supports intelligent and reliable data acquisition to aid efficient monitoring, control, and analysis
• Open for the integration and collaboration with other standards[4]
• Platform-agnostic, unlike its predecessor, which worked only on Windows

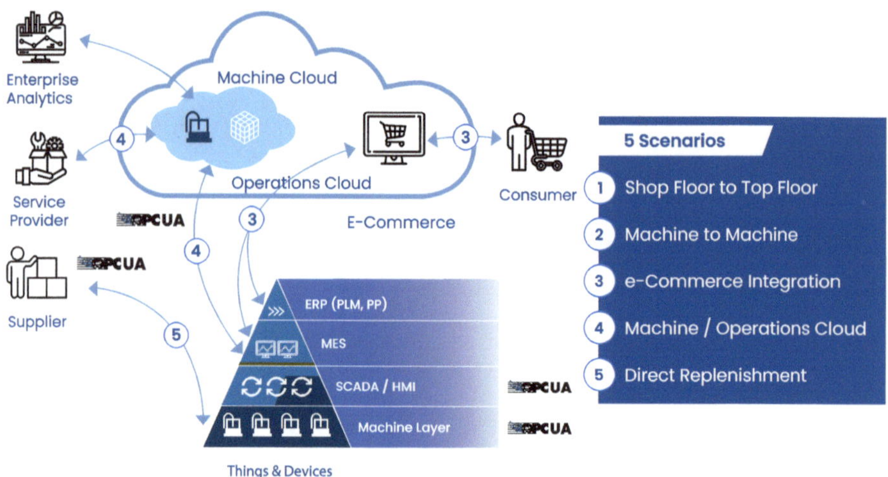

Reference: Illustration #1

Several enterprises have adopted OPC Classic, the predecessor of OPC UA. However, its limitations have made OPC Classic unsuitable for today's cyber-aware world. Vendors have already started ceasing support for this protocol, which means that enterprises that refuse to upgrade to OPC UA are sailing a sinking ship.

This book is your guide to OPC UA and its potential to transform your business. In the coming pages, we will explore the challenges faced by key stakeholders in industrial automation and how upgrading to the OPC UA specification can enable new possibilities.

• • •

#@#%$*
^%@#...
Production Manager
QA Manager
CEO
Supervisor
Reliability Manager

The Stakeholders

Strategy, uptime, and production are the key focus areas in any industrial setup to ensure the business runs efficiently. Strategic insights make the business future-proof. Ensuring uptime and production optimization helps meet business goals. Various stakeholders in the business come together to ensure performance across these focus areas. However, their job isn't easy. Let's look at some of these stakeholders and the challenges they face.

How can I modernize systems and bring-in digital capabilities to improve operations without business disruption ?

Theresa Smith
CDO/CTO/CIO

Theresa is the CTO of a billion-dollar business – Productzon Corp. - in the industrial sector. Her key role is to improve operations and business performance by adopting new technologies. The challenge? Introduce these technologies into the setup without business disruption.

Here's where things get complicated for Theresa. The business has grown organically and inorganically over many years. Every site is at a different level of maturity, using a wide variety of systems and equipment. Strategic acquisitions have added to this complexity with plants that use a completely different set of assets and supervising products. The worst part? It's expensive to make these systems communicate with each other as they lack a common language among them.

Theresa needs to create a technology adoption blueprint that is specific to the unique needs of each plant or each site while being in alignment with the overall business goals. And then, she needs to deploy technology in a way that doesn't affect production. If the business is disrupted, it may end up costing more than the digitalization ROI!

How can I ensure safe, glitch-free operations with minimal disruption?

Steve Staddon
Site Supervisor

Steven is the Site Supervisor at one of the plants of Productzon Corp. His job is to make sure the plant is running efficiently, and the operations are on schedule to meet target production numbers within the allocated budget. The challenge? Ensure the safety and security of the personnel and the site while minimizing operation disruptions.

His job requires Steve to know everything that is going on in the plant at any point in time. Now that's a tall order! He needs eyes everywhere. Are the input materials coming on time? What is the rate of production? Is all the equipment functioning the way it should? Has the maintenance work finished on time? Is there an alarm triggering somewhere that needs attention? Is there a spill, fire, or malfunction that could be hazardous to personnel on the job?

To keep an eye on all these moving pieces and ensure the plant is running smoothly and safely, Steve needs data in real-time. He also needs this data for management reporting. Getting this data from disparate systems using different field communication protocols is Steve's biggest challenge.

How can I make the right decisions
to optimize production ?

Matthew Peterson
Production/Operations Manager

Matthew is the Production Manager at one of the sites of Productzon Corp. His goal? Optimize the plant production to reduce wastage and lower operating costs.

To do this, Matthew needs to have a control on, and handle all production-related activities, equipment, and processes. Is there a mechanism to record and display the work-in-progress inventory and increasing handling costs? Are there any bottlenecks in the production process that are grinding it to a halt? Do people need training to do their job better?

To make the right decisions, Matthew needs data and insights in real-time that could help him decide what he needs to optimize. Just like Steve, getting his hands on this data without jumping through hoops is Matthew's big roadblock.

In addition, Matthew also needs to ensure this data is fed to the cloud without distortion and with the required context so that the top management can access it for predictive analysis.

How can I ensure
first-time-right quality?

Sarah Strasbourg
QA Manager

Sarah is the Quality Assurance Manager at Productzon Corp. and is in a constant battle to maintain consistent and expected quality levels on the items they produce.

Sarah is responsible for ensuring first-time-right quality, improving first-pass yield, and ensuring adherence to industry specifications, standards, and regulations. In addition, she also needs to make sure that the products are shipped on time, delivered without damage, and that customer concerns are addressed rapidly.

To maintain the expected quality standards, Sarah needs to proactively know where the issues are and address them appropriately. What is causing the quality to drop? Where are the defects cropping up? Does the team need training to handle a particular process or machine better? And what process or equipment changes need to be made to improve quality.

She needs a bird's eye view of the entire setup with the ability to drill down to details of what's happening at a very granular level. Unfortunately, at present, she doesn't have a single-pane-of-glass view of the operations and has to manually reconcile several data formats to make decisions. These delays in decisions are adversely affecting quality and impacting throughput.

Sarah also has a key role to play in the Digital Twin initiative adopted by the company. As a domain expert of the plant, she has to find the best possible way to model the plant assets into the cloud environment and get the required metadata from assets to the cloud.

How can I minimize
unexpected downtime?

Ralf Watson
Maintenance Manager

Ralf is the Maintenance Manager at Productzon Corp. and works with Steve, Matthew, and Sarah to ensure there is no unexpected downtime.

His job is to keep the facility running without any hiccups - prevent breakdowns and avoid production delays. He is responsible for managing the installation of new machinery or equipment, repair and upkeep of the facilities, ordering new supplies, and developing maintenance procedures. This requires him to proactively maintain all devices, machines, tools, software tools, network infrastructure, etc. He needs to coordinate with multiple teams to align plans and ensure maintenance is done on time and in time for minimal business disruption. In addition, he also needs to track expenses, keep logs, report progress, assign schedules to workers doing maintenance activities, and ensure that safety procedures are followed.

If Ralf fails to prevent breakdowns, it could cost the company millions of dollars in lost revenues. That's a big responsibility. And guess what he needs to deliver on this ask? You've got it! He needs timely data from every part of the facility and from other teams. Does he get it? That remains to be seen.

How can I increase equipment uptime and reliability?

Andy Sharma
Reliability Manager

Andy is the Reliability Manager at Productzon Corp., and his job is to improve the uptime/availability and reliability of equipment. He works closely with Ralf to ensure proactive and planned equipment servicing and repair before breakdown.

Andy knows that scheduled preventive maintenance can do only so much. He wants to move beyond that to really understand how critical assets at the plant behave – what are their trigger points and failures? He knows getting this information in real-time can help him customize asset maintenance and only schedule it when needed. Moving from schedule-based to need-based predictive maintenance can save a lot of valuable time and maintenance costs. Getting granular insights from real-time, high-frequency machine data can also help him discover and resolve the root causes of failure.

However, disparate systems using a wide variety of sensors, software, monitoring devices, and other technology makes it near impossible to get the data he wants. There is no integration! Unable to access the data he needs; Andy often grapples with issues when it comes to maintaining uptime and reliability.

SCALABILITY
INTEROPERABILITY
INTELLIGENT DATA
FUNCTIONALITY
SECURITY
RELIABILITY
OPC UA

OPC UA Myth Busting #1

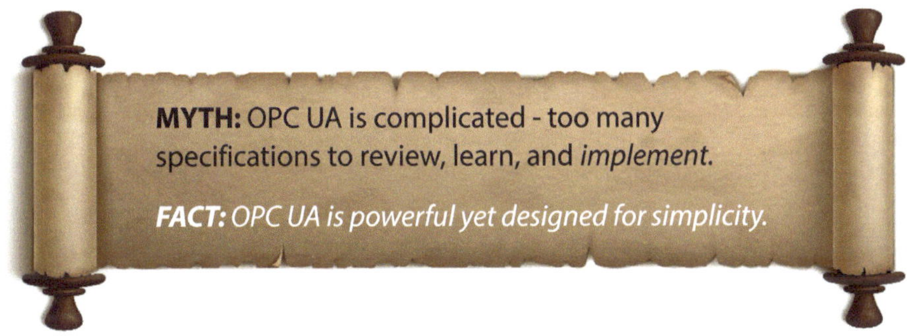

Several factors have contributed to the perception of OPC UA as a complicated architecture. One key thing that has built this perception is the amount of information available. Unlike the OPC classic (D.A, H.D.A, and A.E), which came with mainly three specifications, OPC UA has sixteen main specifications, and the number keeps changing. That's a daunting task to read through. The real question, though, is, "Does the number of specifications matter when it comes to the simplicity of implementation and adoption?"

From a design standpoint - given the functionality, security, reliability, and high performance delivered - the OPC Foundation has really simplified things in OPC UA. Take, for instance, the number of services. OPC UA has less than 40 services to access data, events, and rich information models. In contrast, OPC Classic, for a much-limited functionality, had nearly 70 methods.[5]

Since a single OPC UA Server can implement the functionality of Current Values (D.A), Alarm and Events (A.E.), and Historical Data Access (H.D.A), it is easier to implement and commission it.

So, for example, if an operator is looking to monitor the health of a pump at a site, they would want to know: 1) Current inlet and outlet pressure, 2) Any last events or alarms generated on the pump, and 3) A small sample data trend for the last 20-30 minutes timeframe. To give them this information, OPC Classic would require three different OPC Servers (D.A, A.E, and H.DA) and their respective clients. However, in OPC UA, these three data points can be exported via a single OPC UA Server and a single OPC U.A Client.

Even migrating to OPC UA from OPC classic is easy with the wrappers and proxies provided by the OPC Foundation. In addition, its generic and extensible design makes it easy to develop iteratively and add features over time.

So, while it may seem complicated and painful, configuring OPC UA and managing this change in your business is much simpler. As you start using it and get familiar with the workings, you'll find that the "too complicated" impression was just a myth!

This book will simplify your journey to OPC UA and showcase how it can deliver high-impact outcomes for your business.

● ● ●

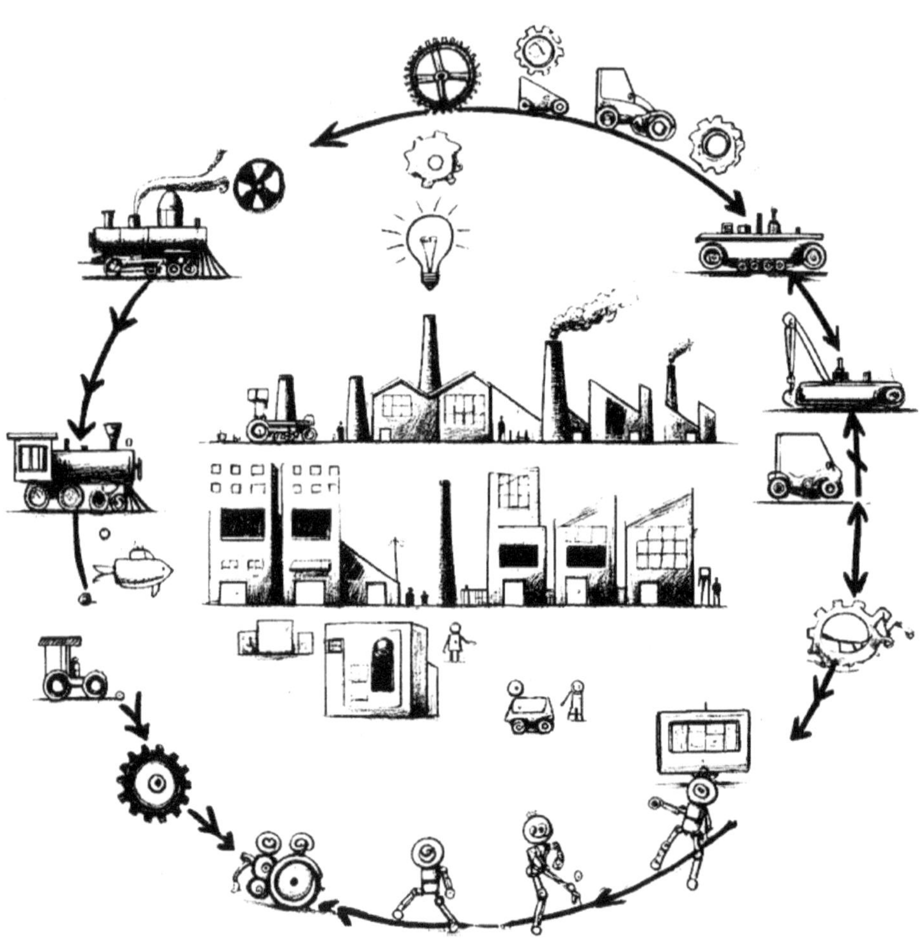

The Landscape

THE INDUSTRIAL AUTOMATION TIMELINE

In the history of production, the focus has always been to increase the profitability by increasing the production, reducing the manual effort and hence the overall cost. Be it the wheel or the first valve in the Roman empire to create water redirection systems, inventors throughout history have brought in systems that made production easier, more efficient, and less dependent on human effort. As pioneers like Henry Ford started deploying scientific methods to assembly lines, the impact was visible to everyone. A moving assembly line helped Ford reduce the time to build one car from 12 hours to just 1 hour 33 minutes! [1]

From the late 18th century, the journey of industrial automation has seen a rapid evolution from pneumatic mechanical systems to digital systems and from siloed digital systems to connected systems. The foundations of Industrial automation were set in the 20th century when Dick Morley developed the first programmable logic controller, and Honeywell created the distributed control systems.

With more technology coming in and autonomous guided vehicles (AGVs) and robots becoming a part of the digital workforce at production facilities, there has been a surge in device data. These cyber-physical systems are the future of industrial automation and are pushing the case for integration and interoperability.

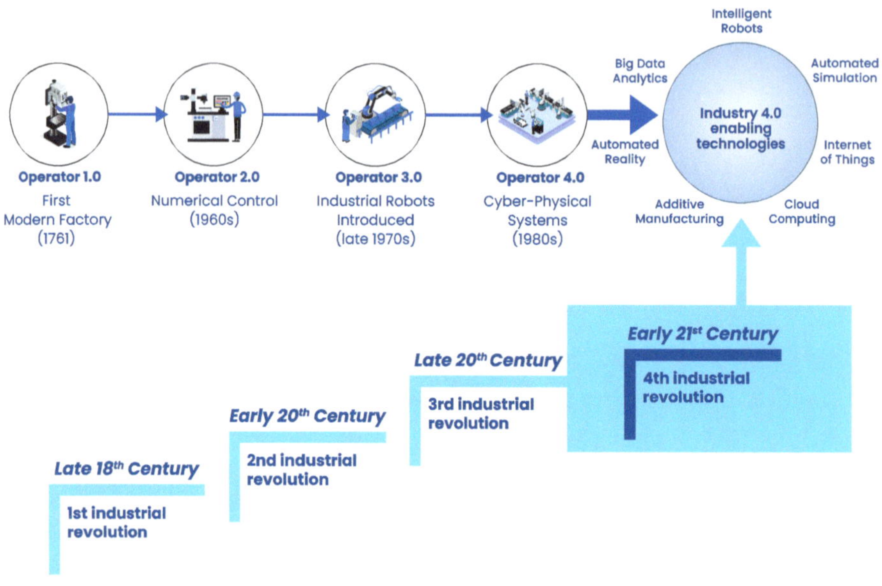

Reference: Illustration #2

THE STATE OF INDUSTRIAL AUTOMATION AT A GLANCE

Industry players are betting big on industrial automation, and investments are on the rise:

- McKinsey found that companies will park 25% of their capital spending over the next five years for investments in automated systems.[2]

- Market.us projects that the Industrial Automation Market will reach USD 493 billion by 2032, growing at a 9% CAGR between 2023 and 2032.[3]

Investing money in industrial automation systems makes sense because of the phenomenal ROI they can deliver:

- Increase productivity by up to 30% and create significant cost savings.[4]

- Improve safety in the workplace, reducing the number of work-related injuries and fatalities by up to 50%[5]

- Reduce production costs in oil and gas by up to 20%[6]

And those that can drive interoperability between the various systems will come out winners.

- Companies with high interoperability grew revenue 6X faster than their peers with low interoperability.[7]

However, challenges remain when it comes to the adoption of industrial automation [8]:

- 61% say the company's general lack of experience with automation is a roadblock

- Implementing a single set of solutions backed by integrated and interoperable programming and platforms is a challenge in the jumble of legacy technologies

- Potential inability of machines to interface with products

- Concerns about safety and cyberattacks

ADOPTION DRIVERS AND STUMBLING BLOCKS

Did you know that the original Ford Model T in 1908 cost more than the all-new 2022 Ford Maverick? Its sticker price of $850, adjusted to today's consumer price index, would be $25,223.10. In contrast, the Maverick has a starting price of just $19,995. So essentially, we are paying less for a car with more features than we would have over a century ago.

Automation has been one of the key factors that democratized consumption by enabling mass production that made everything so affordable. But at the same time, the need to produce price-competitive products has put severe pressure on the operating margins. Industrial companies must find ways to curb their cost of operations and stay profitable.

This is not an isolated case in the automotive industry alone. Take, for example, oil & gas (O&G), where the rise of renewable and alternative fuel sources and consumer sentiment shifting to these sustainable sources has changed business dynamics. The O&G players now need to price competitively and must optimize the cost of production to keep prices in line with market expectations.

Consumer behavior and market trends are the two drivers of progress in any industry. Today, connected global marketplaces, diminishing barriers to entry, and changing business and consumer dynamics have brought in a lot of volatility in the market.

To stay competitive and productive in this new consumer and market landscape, industry players are turning to automation. There has been a drastic shift in the way automation is viewed – from something that helped them operate a plant safely and could be run in isolation to smart systems that optimize all aspects of production. The complexity of operations has changed multi-fold over the decades. Despite the complexity, the ask remains to be productive and competitive, creating the need for complex automation. To keep up with this ask, automation systems have also evolved - from geometric automation to PLC automation, to DCS automation, to now connected automation that generates massive volumes of data.

To support this automation and make their business future-ready, companies also need to modernize legacy tech and bring in new technologies. Today most plants and facilities are running legacy assets and systems that are not compliant with current evolving standards. These legacy systems are also difficult to connect and are incapable of handling the data volumes generated by modern plants. As most of the technologies in use are outdated, they also create security risks, especially with increasing cyber threats. Large data volumes passing through these systems create issues of their own – reliability, security, scalability, audit, etc.

To align with the continuous evolution of automation, industry players are adopting new technologies that can evolve with moving standards and sustain business needs for the coming few decades. However, that's not an easy choice to make.

Essentially, industrial companies are grappling with three key issues:

1. How do we modernize legacy technology with zero business disruption?

2. How do we control costs to stay profitable and minimize the impact on our bottom line?

3. How do we adopt newer technologies to future-proof business for the next few decades?

One of the keys to answering these questions lies in data. In the coming pages, we will discover how OPC UA is the silver bullet to handle this growing complexity and solve the most pressing problems of industrial automation.

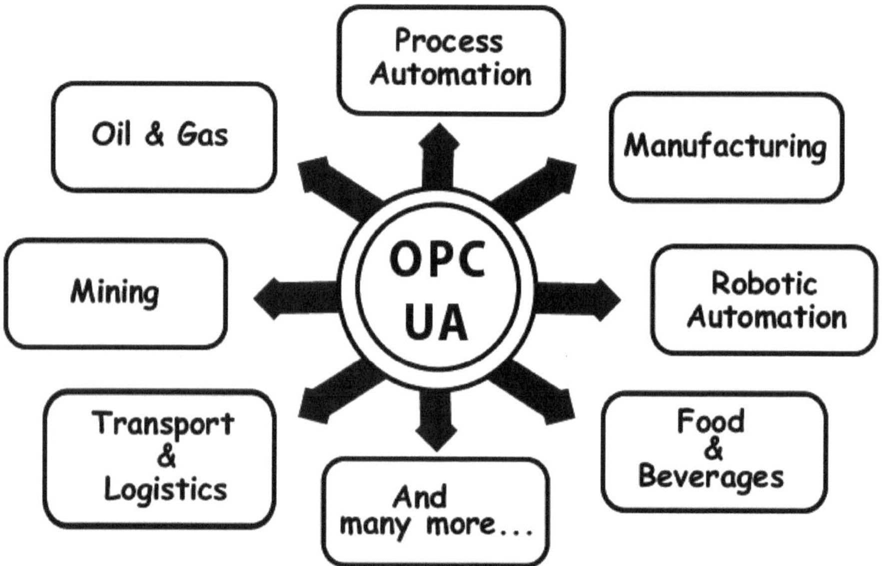

Process
Automation
Oil & Gas
Manufacturing
OPC
UA
Mining
Robotic
Automation
Transport
&
Logistics
Food
&
Beverages
And
many more...

OPC UA Myth Busting #2

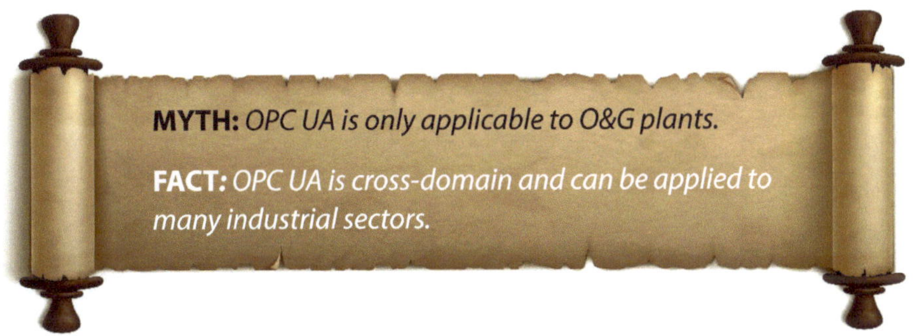

The myth that OPC UA is only applicable to O&G companies has its roots in historical usage. In the industrial sector, O&G players were some of the first to deploy automation. They were also the first adopters of OPC Classic as the criticality of their business demanded rapid action. For instance, any glitch or malfunction that can grind production to a halt is still okay in an industry like automotive. It may delay time to market but won't destroy billions of dollars' worth of inventory. On the other hand, stopping production in an oil rig is not an option. Unlike other industries, the nature of output is continuous. And so, they heavily adopted automation to monitor and keep the production always on. In fact, this need, for O&G industries to have systems that talked to each other, gave rise to OPC Classic.

This heavy usage by the industry created the impression that OPC was only applicable to Oil & Gas. That, however, is a myth. The OPC Classic, and now OPC UA, are both designed for application across industries and are successfully being used by companies in various domains.

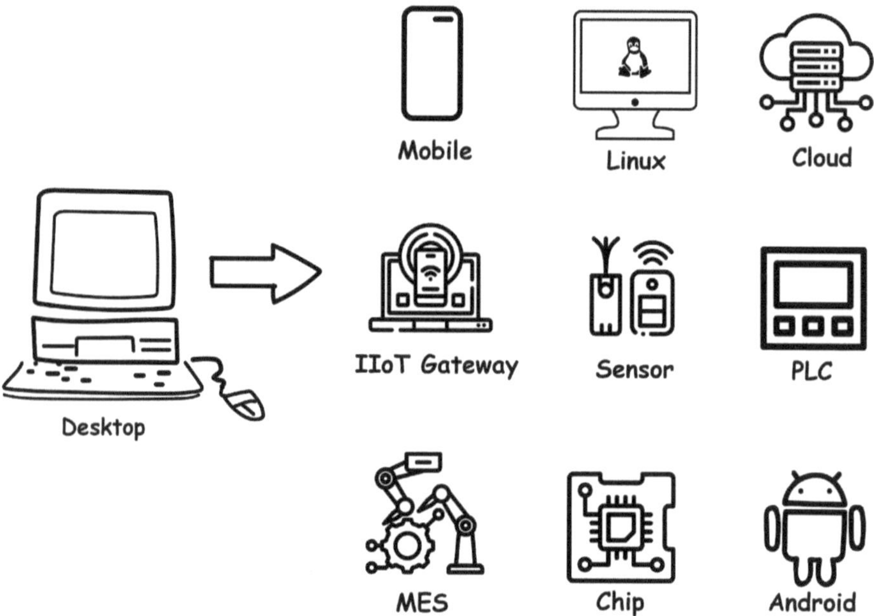
Mobile
Linux
Cloud
Desktop
IIoT Gateway
Sensor
PLC
MES
Chip
Android

The Evolution of OPC UA

Before the OPC era, industrial operations relied on multiple devices from various vendors supporting different protocols and proprietary software components. As illustrated in the figure below, each hardware component was tightly coupled with visualization and control systems. Any addition of new devices needed an upgrade of the visualization and control software or rigorous training for the operators on proprietary tools. The industry sorely felt the need for a standard interface, and OPC Classic was born.

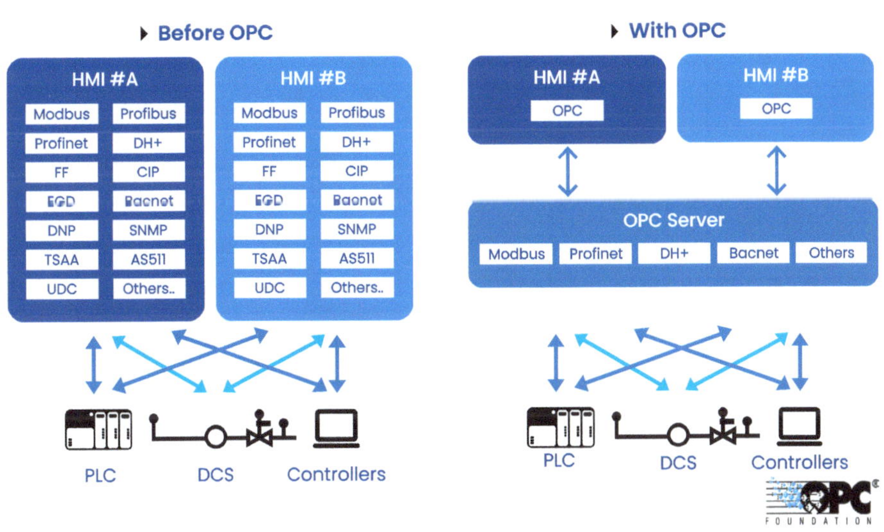

Reference: Illustration #3

During those days, Microsoft COM/DCOM (Component Object Model) was the most in-vogue technology for sharing information between systems. It made sense for Industrial Equipment Manufacturers and Microsoft to come together and co-create something. The specification they came up with was called OPC or 'OLE for Process Control' and was exclusively tied to the Windows operating system.

As a specification solving the most common problem in industrial operations, OPC DA saw rapid adoption and soon became the most accepted solution for interoperability between field devices and HMI applications. In the meanwhile, to support non-windows platforms, OPC XML-DA specification was introduced, which used XML as the basis of information exchange over HTTP/SOAP and web service technologies.

However, with increased adoption, the limitations of OPC DA became more prominent. And due to high resource requirements for processing the XML, OPC XML-DA found limited adoption. There was a need for something that unified everything – a platform-independent standard that could possibly be adopted even in industries beyond manufacturing. This intention gave birth to OPC UA.

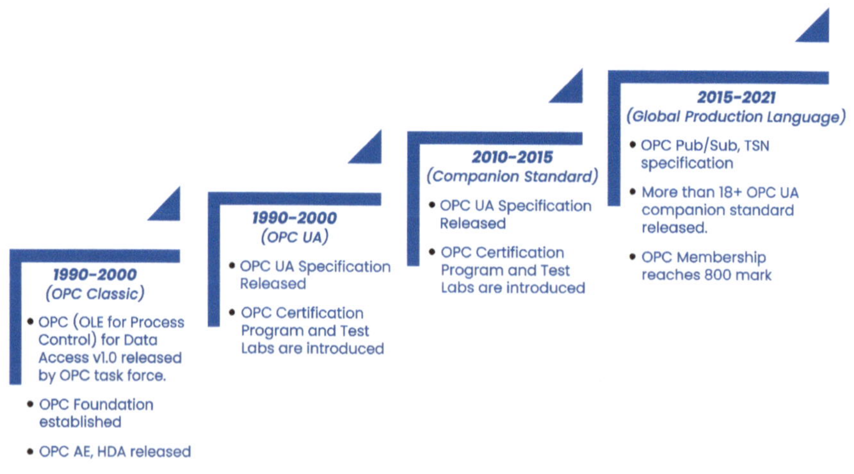

Reference: Illustration #4

OPC CLASSIC Vs. OPC UA

Criteria	OPC Classic	OPC UA
Communication Protocol	DCOM (Distributed Component Object Model), limited to Windows platforms.	Platform-independent, uses TCP/IP, UDP and MQTT, or messaging services, enabling communication with Linux, macOS, etc.
Network Security	Relies on DCOM, which can be challenging to configure securely.	Comes with built-in security measures such as encryption, authentication, and authorization.
Scalability	Limited due to dependence on DCOM.	High scalability, can efficiently connect to a larger number of devices across firewalls and networks.
Data Representation	Uses Data Access (DA), Alarm & Events (A&E), and Historical Data Access (HDA) separately.	Unified data model that combines all features of OPC Classic and includes methods and programs, allowing complex data types and object-oriented data representation.
Interoperability	Limited to Windows-based systems, potential compatibility issues with different Windows versions.	High interoperability due to platform independence, works with numerous systems, and supports cross-platform operability.
Real-time Data Access	Provides real-time data access but may face latency issues due to DCOM.	Efficient real-time data access with better performance across networks and firewalls.

Redundancy	Does not natively support redundancy.	Supports redundancy, leading to better reliability and system stability.
Loose Coupling	Tightly coupled because of dependency on DCOM and Windows platform.	Offers loose coupling due to its platform independence and communication over Web services.
Information Modelling	Limited, mostly flat address space and simple data structures.	Allows complex information modeling with hierarchical address spaces and object-oriented structures.
Service-Oriented Architecture	Does not natively support SOA.	Designed around SOA, providing more flexibility for developing scalable, secure, and reliable services.
Platform Independence	Limited to Windows platform.	True platform independence - can operate on any system that can run a Web service.
Outcomes	Limited outcomes due to restrictions of DCOM and Windows-only operability.	Achieves better outcomes due to versatility, security, and wide interoperability.

DEMOCRATIZING INDUSTRIAL OPERATIONS WITH OPC UA

Operations in industrial plants and facilities are governed by different layers of technology that enable information flow from the field level all the way up to the management level. Until about a decade or so ago, these layers of sensors & signals, controls (PLC, PC, PID), supervision

(SCADA, HMI), planning (MES), and management (ERP) were rigid, and vendor locked. Companies had to invest in expensive systems to enable the flow of information, and maintaining and upgrading these devices was cumbersome.

The evolution of OPC UA brought in the full mesh network architecture that removed the vendor lock-in and dependence on costly, closed systems. In a full mesh architecture, data flows both ways, and any node can talk to any other node. Using OPC UA, any asset in the plant can connect with any other asset or system quickly and securely, and the technology dependency is reduced. The layers fall away in favor of a networked and connected smart factory.

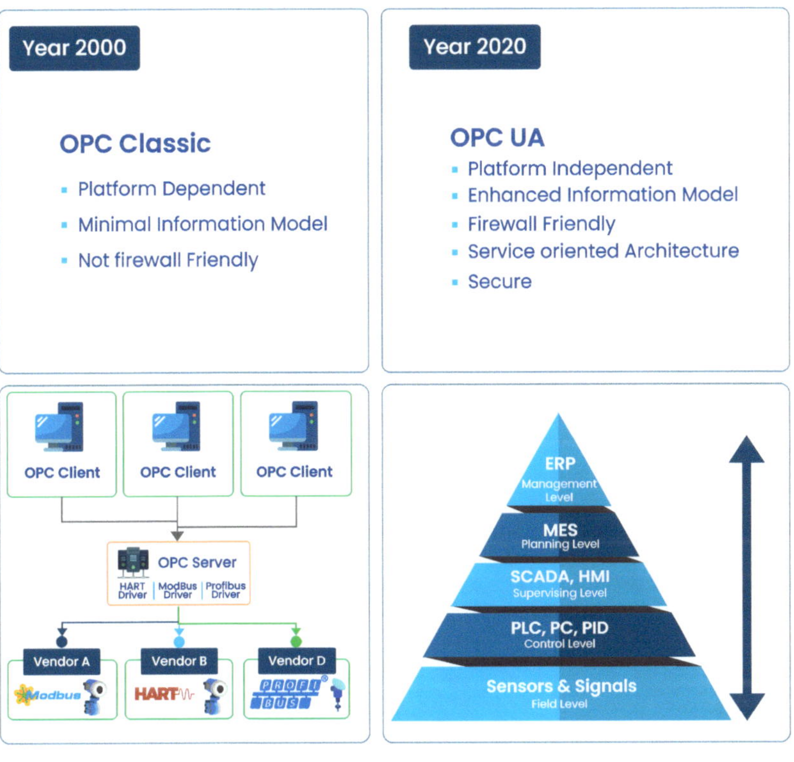

Reference: Illustration #5

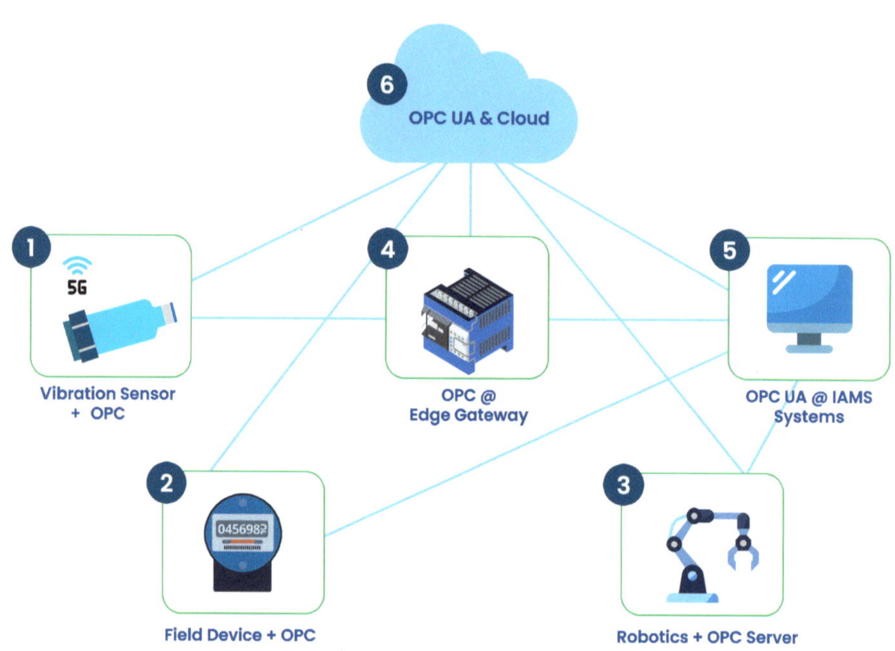

Illustration: OPC UA Enabled Full Mesh Network - Democratizing Field Data

One of the key challenges in the industry is that sensors are tightly coupled with IO-Modules or Programmable Logic Controllers (PLCs) and have multiple communication protocols and network zones. It is challenging to integrate a large number of distributed devices, and only a fraction of information can be transmitted sporadically.

With OPC UA, all devices and systems in a plant – from the smallest sensor; to the most complex equipment. It ensures that all the interconnected units speak the same language and overcome the communication issues that otherwise arise from various proprietary systems. For instance, it can collect data from a sophisticated robot and move it to the cloud, OR; can connect devices in the field over 5G via edge gateways that bring device data to the IMS. This kind of interoperability was not possible earlier when all assets and devices were speaking their own set of languages.

OPC UA enables this connectivity by using different profiles (Nano, embedded, standard server, etc.) to customize the capabilities/ feature set to the device it needs to be deployed on. This customization makes it easier to get information from both small devices like sensors and larger devices such as Distributed Control Systems (DCS) and PLCs. For example, for sensors; OPC UA will deploy the nano profile and only incorporate basic features taking up lesser memory and bandwidth while enabling the required information flow. As the device size grows, embedded or standard profiles with more features can be deployed at the company's discretion.

As OPC UA started exposing data via standard objects and allowed for expansion, the other major players from the industry (protocol and specifications like BACnet and ISA95) built companion specifications to expose their Asset Model data via OPC UA. This meant that the protocol devices could have a 1:1 mapping of their existing Asset Model with OPC UA conveying the same meaning without changing context. That was a disruptive innovation and opened exciting possibilities. This collaboration also paved a path for OPC UA to become a universal protocol in the near future. Now that various devices with different components and footprints could communicate in OPC UA, it further democratized the choice of assets in a plant or facility and made it easier to monitor assets and their condition.

How OPC UA Helps Industrial Operations		
Enables quick and secure communication across devices and systems.	Removes vendor lock-in.	Enables data flow from the edge to the cloud with no loss of information.
Removes the need for expensive systems to interpret device data.	Reduces asset management and troubleshooting costs.	Gives companies the flexibility to choose components for their plants and facilities.

The richness of the OPC UA information model also made it possible to bring robotics into the fold by transferring not just the data but complete semantic information about the data. The OPC UA information model can be extended to include companion information models with other industry standards and vendor-specific information models. Today, using a vast library of objects, any robotics company can represent its data using OPC UA. OPC UA also enables another important aspect of robotics data - bringing deterministic behavior. This is done via the OPC UA FX initiative that will enable OPC UA data to be sent over a TSN (Time-sensitive Network).

While ensuring connectivity, OPC UA also ensures that no information is lost when data moves from the edge gateway to the cloud. As there is one protocol and the same standards, there is almost zero loss of information.

And finally, OPC UA also makes a significant impact on the bottom line by reducing the cost of managing software assets. Instead of maintaining several teams for each software product onsite to troubleshoot, companies can now rest assured with just an OPC UA team to manage almost all their assets and ensure uptime.

"OPC technology has become a de facto global standard for moving data from industrial controls to visualization up to MES/ERP and IT cloud levels. The rapid expansion of OPC UA in automation, IIoT, and into new, non-industrial markets suggests that OPC will remain an important technology for multivendor secured interoperability, plant floor-to-enterprise information integration, and a host of other applications yet to be envisioned."

– Craig Resnick, Vice President, ARC Advisory Group[i]

SECURE
OPC UA

OPC UA Myth Busting #3

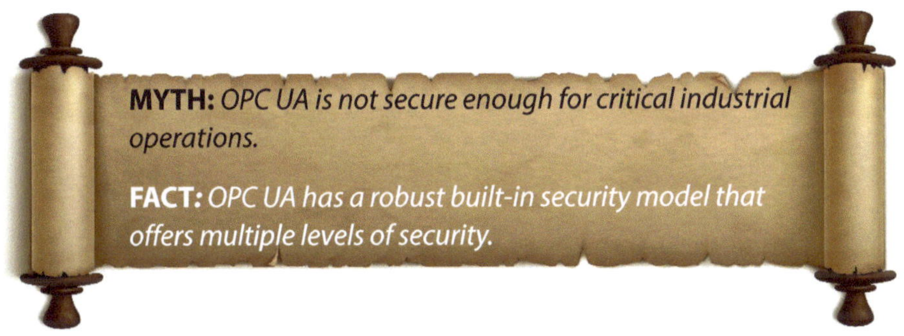

The myth that OPC UA is not secure enough for critical industrial operations stems from a variety of misconceptions and misunderstandings. OPC UA has extensive security capabilities which may not be understood by all, leading to doubts about its effectiveness. In addition, misinformation, biased perspectives, and confusion with older versions may have contributed to the mistaken belief about OPC UA security.

Like any system connected to a network, OPC UA is potentially vulnerable to attacks. OPC UA's security measures must be implemented correctly to be effective. Some reported security incidents might have contributed to the perception that OPC UA is not secure enough. However, these incidents often involve poorly implemented security measures rather than intrinsic weaknesses in the protocol itself.

The fact is that OPC UA comes with a robust built-in security model that provides multiple levels of security specifically tailored to address the challenges inherent in securing industrial systems. OPC UA implements security at the application layer, thereby extending protection to the actual data and not just the communication transport. This aspect is critical for securing sensitive industrial data that needs to remain confidential and intact. With user authentication mechanisms and controls, strong encryption methods, and industry-leading encryption standards to guard against eavesdropping, tampering, or message forgery, and data integrity validation OPC UA is the safest protocol around.

IoT/IIoT/AIoT
Cyber Security
Edge Computing
Digital Twin
IT-OT Integration
Scalability
OPC UA

Enabling New Possibilities

Productzon Corp., a thriving automotive company, was navigating the challenges of the modern industrial landscape. The company was known for its pursuit of innovation and efficiency to stay ahead of the game in a competitive sector.

As the CTO of Productzon Corp., Theresa always had an eye on the horizon. She knew that profitability hinged on their ability to adapt and harness cutting-edge technologies. She wanted to ensure her company was able to leverage the massive volumes of data it generated to make better business decisions. To ensure the systems talked to each other and the information could be easily understood by the users, she championed the adoption of OPC UA across Productzon Corp's plants and facilities. Theresa was happy to note that within a short timeframe, the unified data model was delivering significant operational improvements. Data discrepancies had reduced, and the enhanced capability to network with multiple devices and platforms reduced system integration costs.

OPC UA was also making life easier for site supervisor Steve. Early every morning, Steve would analyze real-time data from the production floor. He appreciated how OPC UA facilitated the seamless integration of different systems and devices, significantly reducing system compatibility issues. With an easy-to-understand interface, the metrics offered by OPC UA helped him implement site improvements. For example, getting accurate energy consumption data helped him implement energy-saving strategies, leading to a significant reduction in energy costs while contributing to Productzon Corp's sustainability goals.

In the bustling production line, Matthew, the production manager, was floored by how OPC UA improved the reliability of the plant operations. With high availability and failover capabilities, paired with an Alarm Management Software exporting data via OPC UA, it eliminated unplanned downtime. By monitoring real-time data on machine performance, he could also anticipate potential issues and act proactively, reducing process disruptions.

Back in her office, Sarah, the Quality Assurance manager, analyzed the recent reports. She observed improvements in data integrity, resulting in a higher quality of products. Additionally, the system's seamless compatibility with advanced quality control mechanisms enabled her to respond quickly to any anomalies, enhancing customer satisfaction rates.

One day, the maintenance manager, Ralf, received an alert from the OPC UA-enabled predictive maintenance system. An inspection was due for a piece of equipment showing signs of wear. With this advanced notice, he was able to schedule the maintenance at an off-peak time, reducing unexpected equipment downtime and maintenance costs.

Meanwhile, the reliability manager, Andy, admired how the OPC UA interface usage in the condition monitoring software led to a reduction in equipment failures. The alarm and event data made available to conditioning monitoring software enabled his team to react quickly to process anomalies, contributing significantly to the system's overall reliability.

Every day, Productzon Corp. deals with an enormous amount of sensitive data. Their cybersecurity lead, Alice, valued the built-in security mechanisms of OPC UA. With advanced encryption, user authentication, and a robust defense against cyber threats like DoS attacks, OPC UA has almost eliminated cybersecurity incidents in their plants.

Thanks to OPC UA, Productzon Corp. experienced improved decision-making, lower Total Cost of Ownership (TCO), and a significant reduction in waste. Their capability to scale operations rapidly to meet market demand, coupled with enhanced supply chain collaboration, strengthened their competitive position in the automotive industry.

OPC UA had become an invisible yet indispensable part of Productzon Corp., threading its way through every operation and decision. As the company continues to innovate and adapt, its leaders know that with OPC UA, they're driving not just cars off the assembly line but also their company into a future of unprecedented efficiency and growth.

UNLOCKING THE POWER OF DATA

We've seen how the siloed nature of traditional industrial systems hinders data-driven decision-making. OPC UA breaks down these siloes and makes data accessible to all stakeholders in an organization, not just to specialists or top-tier management. Imagine what possibilities that can unlock! When Matthew, Steve, Sarah, Ralf, and the other stakeholders introduced earlier, get real-time access to industrial data, magic happens. Suddenly the site operations are running smoothly; there are no unplanned downtimes, the equipment and machinery perform better and for longer, wastage and losses reduce, and the business sees a surge in productivity, efficiency, and innovation. All because the devices across the board – from different manufacturers and running on different platforms and protocols - are talking to each other.

This democratization of data in a secure framework is instrumental in transforming the industrial operations landscape to set the stage for a smarter, more efficient, and future-proofed industrial environment.

Illustration: OPC UA-Enabled Business Value Chain

AMPLIFYING OUTCOMES AND INTEGRATIONS WITH OPC UA

Industrial operations are going through a transformation driven by the convergence of several technologies. In this landscape, OPC UA stands as a central pillar that can tie these concepts together.

Powering IIoT and AIoT

The Industrial Internet of Things (IIoT) and the Artificial Intelligence of Things (AIoT) represent a powerful synthesis of technologies. IIoT seamlessly collects, aggregates, and analyses data from a wide array of devices and sensors across an industrial environment data and AIoT ensures that this data can be leveraged to automate decision-making and improve operational efficiencies and business outcomes.

For instance, in a water treatment plant, IIoT devices such as pumps and pH sensors can collect and transmit data about the health of the system and the quality of output. This data, ingested by an AI system, can generate possibly life-saving insights on water contamination alerts and possible equipment failures and even recommend measures such as optimal chemical dosing, etc. However, on-ground, collecting this data across devices is not an easy task. As we discussed before, the reality in industrial operations is that most components don't talk to each other. And that is where OPC UA comes in.

OPC UA, with its ability to standardize communication and data modeling across diverse devices and platforms, simplifies the complex and challenging task of gathering data reliably and securely in real-time. In addition, it also facilitates the integration of AI in industrial operations by providing the infrastructure for a seamless flow of data from IIoT devices to AI systems. With OPC UA's cohesive and unified data sharing, device data gets pushed all the way from the shop floor to the cloud with minimal distortion and without the loss of context.

Enabling Edge Computing

Often industrial operations require the need for data processing closer to the source of data generation. A great example is trucks used for mining. Today the mining industry uses smart trucks equipped to the gills with smart sensors that sense everything!

However, the nature of operations requires that the mining trucks ply in very remote areas, often with limited or no connectivity. This means that the sensor data cannot be uploaded to the cloud in real time, and the trucks are not able to receive insights on this data. So, if there is a breakdown happening or some part is failing, a truck out of the network zone will not get to know it till it's too late. To get the benefit of data-driven insights even in remote locations, the trucks need edge computing to analyze the sensor data locally. And when the truck is back online, this edge device can transmit data and analytics to the cloud.

Similarly, edge computing is vital in operations where low latency is business critical. These operations cannot wait for their devices to connect to the cloud and then get a response back. In these cases, sensors from the facility get connected to an edge computing device which then transmits data when feasible to the enterprise cloud. This allows for efficient data processing, optimizes network resource use, and enables faster insights.

The ability of OPC UA to fit into smaller devices with its micro and nano specs makes it the ideal protocol to integrate edge computing. By offering a platform-independent, service-oriented architecture, OPC UA ensures seamless communication between plant sensors to edge devices to central servers and makes edge computing viable.

Information Modeling for Digital Twin

Digital Twin technology allows companies to create a digital replica of a physical entity, such as plants and facilities, oil rigs, and each individual component of these facilities, such as furnaces, air conditioners, boilers, pumps, etc. A Digital Twin makes it easier to understand and manage complex systems in real time, enabling predictive maintenance, resource optimization, and advanced simulation. For instance, blowouts resulting from failures of pressure control systems are a real threat in the oil and gas industry. And if the wellhead is hot when this uncontrolled release of crude oil happens, it could lead to a catastrophic fire.

A digital twin of the oil well could prevent blowouts by giving timely and actionable information about the temperature and pressure status of the rig components. However, before OPC UA, getting this information was a challenge.

OPC UA has revolutionized Digital Twins technology with rich information modeling that no other protocol before it addressed in such detail. Standardized interfaces and models of OPC UA ensure consistent data exchange and create an effective representation of industrial assets and systems. With OPC UA, instead of flat data values for an asset, you get object representative values, and even the metadata is passed on to the top levels, and operators can easily see the connections between various assets or components that are a part of the same device. In the OPC Classic era, an operator needed to have extensive domain knowledge to make a decision based on the data available. However, OPC UA makes the linkages obvious, simplifying decision-making and reduces people dependency. Earlier this level of asset modelling was locked inside expensive software which is now available with a simple OPC UA client.

Amping-up Cybersecurity

In today's interconnected industrial landscape, cybersecurity is a vital concern especially in critical infrastructure operations like smart grids. With increasing hacker sophistication, cyber-attacks are always a moving window and organizations need to stay a step ahead of attackers. OPC UA, is one of the most secure industrial communication technologies available today and offers protection against unauthorized access, sabotage, modification of process data, and careless operations.

"When it comes to networking and standardization or cybersecurity – the industrial requirements are increasing. Here, the OPC UA communication standard offers the best conditions for digitalization use cases, independent of platform and manufacturer and with integrated security mechanisms."

– Thomas Hahn, Siemens AG, OPC Board Member[i]

Allowing Scalability

Scalability is a crucial factor for industrial operations. As companies grow and their operations evolve, their systems need to adapt and scale accordingly. If systems don't scale in tandem with growth, it creates siloes and the company loses the advantages of Industry 4.0 connectivity.

OPC UA is uniquely suited to meet scalability requirements. Its architecture supports industrial operations ranging from a single machine to complex, global networks and provides mechanisms for data discovery and automatic reconfiguration, supporting growing networks without manual intervention. Moreover, OPC UA supports both vertical and horizontal scalability and can also efficiently manage growing data volumes. It can handle an increase in data load by adding more resources to a single node in a system and can expand capacity by connecting multiple hardware or software entities to work together as a single logical unit.

A real-world example of OPC UA's role in flexibility and ease of expansion can be seen in the deployment of smart grid technologies. As the grid expands, new devices such as smart meters, sensors, and controllers are constantly added. With OPC UA, these new additions can be seamlessly integrated into the grid, ensuring smooth communication and data exchange irrespective of the grid's growing complexity. Similarly, as a manufacturing setup expands, new machinery and systems need to be incorporated without disrupting existing operations. OPC UA's standardization capabilities enable the easy addition of new machines to the line, regardless of the manufacturer or underlying platform.

Facilitating IT-OT Integration

In the early days of industrial operations, the lack of a common communication protocol meant that the plant was isolated from the business network. However, today, discovering and implementing higher-value enterprise-wide use cases require the integration of data from both IT and OT networks.

In the pre-OPC era, if let's say, one wanted to streamline inventory management. Then this activity could only happen at a plant level using the data available at the plant. However, as COVID-19 demonstrated, disruptions can happen any time, and just optimizing inventory at a plant level is not enough. Staying resilient in a turbulent business landscape requires integrated inventory management at an enterprise level. This means integrating data from all parts of business – production, warehouse, sales etc. – that is sitting across IT and OT systems to create a cohesive picture of inventory needs. Moving this data to an application in the cloud can help enterprise business decision makers get key insights in real time. For example, "Given the current inventory levels, what is my production capacity?" or "What is the lowest cycle time to produce XYZ product?" Or "How much quantity of a component should be ordered from the supplier and when should that order be placed to minimize production disruption?"

Integration of IT and OT is crucial for realizing the full potential of the IIoT. OPC UA provides a common language for both IT and OT systems, facilitating seamless data exchange and harmonization. With OPC UA, data can securely pass through the DMZ firewalls on a single port so as not to compromise the firewall.

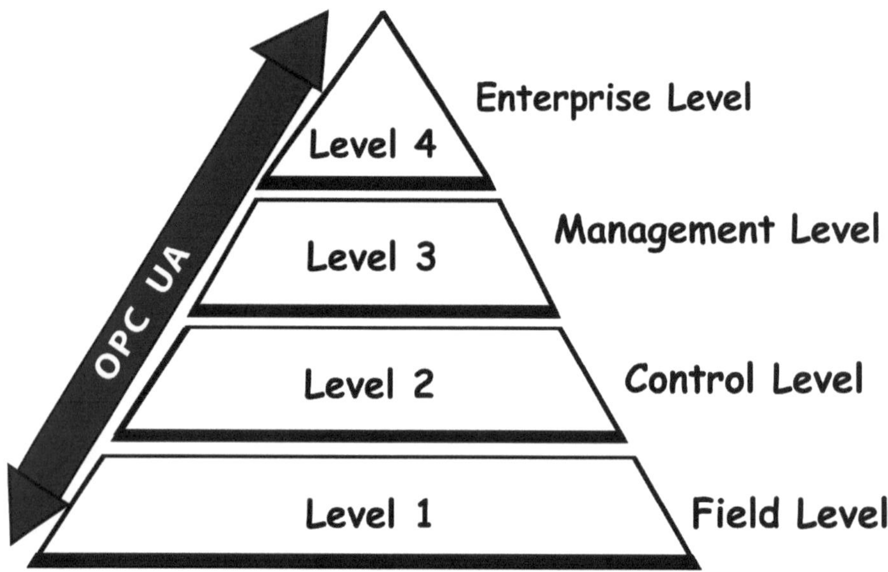

Enterprise Level
Level 4
Management Level
Level 3
OPC UA
Control Level
Level 2
Level 1
Field Level

OPC UA Myth Busting #4

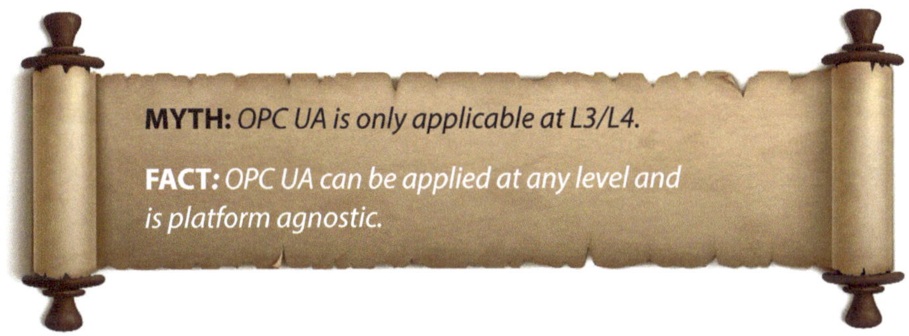

The myth that OPC UA is only applicable to L3/L4 is tied to the evolution of this protocol. The OPC technology was first created using Windows COM/DCOM technology and couldn't be used on a non-Windows product. Embedded products like transmitters and PLCs don't run any Microsoft technology and weren't able to use the first evolution of OPC. The disconnect between the operating systems (OS) of devices and sensors in the factory-made people think that OPC can only be used for more advanced applications.

However, OPC UA has bridged this gap with nano and micro specs, and it can now be deployed even on an embedded node. Moving away from OS dependency and becoming platform agnostic has made OPC UA a preferred interoperability standard, especially for new products entering the market. This preference is only boosted by the ironclad security it offers. In the era of increasing industrial connectivity that requires high data visibility, security is non-negotiable. To enable secure IT-OT integration, OPC UA is the most optimal technology.

OPC
UA
Success

OPC UA Success Stories

Several organizations have reaped the benefits of OPC UA, connecting their operations and enabling insight-led decision-making. The case studies in this section demonstrate the transformative impact of OPC UA technologies across various sectors, including manufacturing and energy. By implementing this standard, organizations have streamlined their operational workflows and gained real-time insights into their production processes and supply chains. This enhanced connectivity has enabled more informed decisions, improved operational efficiency, and significantly reduced costs.

Additionally, OPC UA's compatibility with existing systems and its robust security features have positioned it as the preferred choice for organizations aiming to modernize their infrastructure and leverage the Industrial Internet of Things (IIoT) capabilities. These success stories highlight how OPC UA has become an integral part of the digital transformation strategies of these progressive organizations.

Case Study 1:
A MULTINATIONAL OIL & GAS COMPANY GETS A UNIFIED VIEW OF THEIR ASSETS

A multinational oil & gas company had multiple isolated systems, spread across locations, and running on different protocols. Each location had their own dashboards to monitor these systems, but there was no unified view. To get a better understanding of this data and analyze it to create value, the company wanted to move it to the cloud. However, getting all that data – from a variety of systems and of diverse data types - together onto the cloud without security breaches and loss of data was not feasible with the existing conditions.

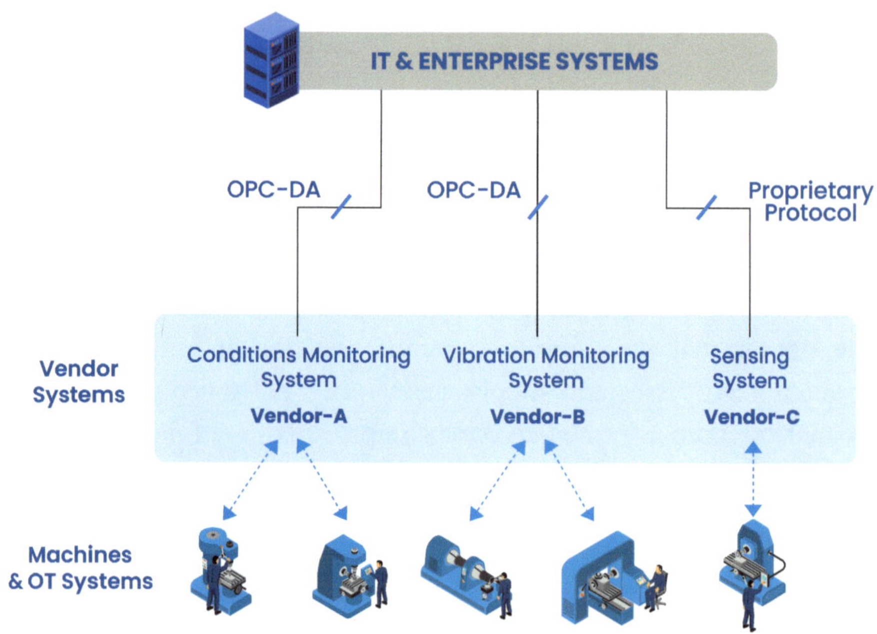

Illustration: Exsiting Brown Field Setup

The company had multiple instances of asset management systems collecting data and multiple instances of anomaly detection software in the same OT network working in isolation – all provided by different vendors. The existing OPC Classic servers were not equipped to cross the DMZ firewall. With multi-generation systems in the picture, it was difficult for the customer to aggregate and normalize this data and move it to the cloud.

To solve this challenge and cross the firewall safely without opening a lot of ports, a native OPC UA server was built on top of the customer's asset management software using their native APIs. Where they were running the OPC Classic servers, an OPC UA wrapper was used to convert the classic data to UA and facilitate transfer across DMZ. The OT data in OPC UA TCP/IP format was moved till the DMZ level. Next, the Utthunga uOPC Data Bridge was added at DMZ level to pull this data from OT and push it to the cloud using MQTT broker. Since this data was in OPC UA Pub-Sub format, it was supported by most of the software in the IT/Cloud network.

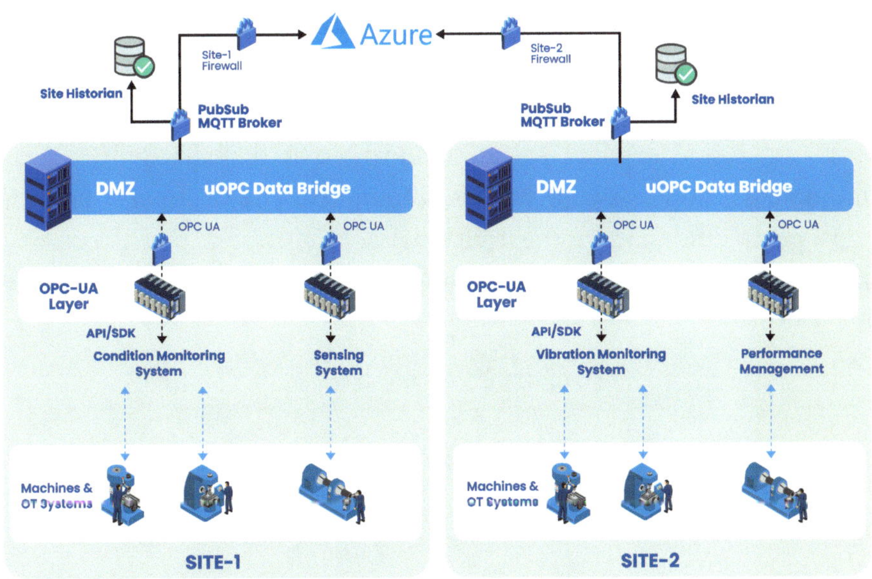

Illustration: OPC UA-Enabled OT Data Network

Case Study 2:

THE JOURNEY TO A DIGITAL FACTORY FOR A STEEL PLANT

A major steel plant's current OT setup was done over a period of time and had many limitations. For instance, too many legacy protocols and legacy devices in the setup were causing frequent breakdowns. Old systems were also difficult to maintain and repair. In addition, they had three types of data getting stored in their persistent storage. These were:

1. Structured data which was the Monitoring data from PLCs and Sensors.

2. Offline data which was stored in the XML files and uploaded manually.

3. Video streaming data for vehicle movement of raw goods.

Legacy protocols were slow and inadequate to meet digital business needs. For instance, instead of a couple of seconds, a report took up to three minutes to load! Imagine the delay and frustration! In addition, poor design and product selection meant that they needed to clean and manage the Persistent storage frequently. The setup also had too many ports open and connectivity between different products was a challenge due to different data formats.

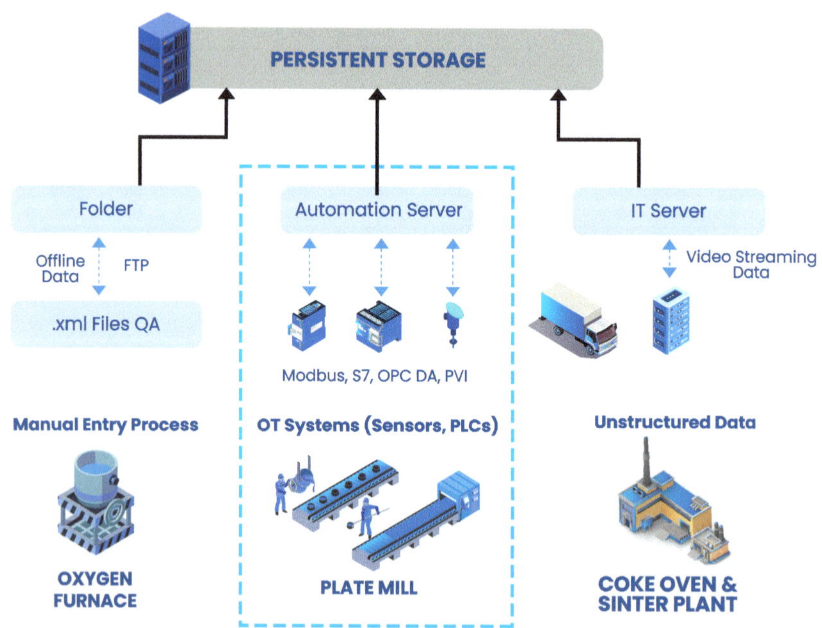

Illustration: Existing Setup of a Steel Plant

To begin their transformation to a digital factory they first wanted to clean up the OT setup responsible for managing structured data (shown in center). The goal was to:

1. Create a cybersecure OT architecture with security level at SL2 from ANSI/ISA-62443-4-2.

2. Clean up legacy devices and protocols and create options to build connectivity across them.

3. Ensure no data loss or distortion occurs while moving data to the cloud.

To redesigning this OT setup, the outdated/unsupported protocols were cleaned up and replaced with the standard protocols. The Field Protocols were limited to L1-L2 levels and deployed OPC UA TCP\IP version in higher levels. The OPC Classic Protocols were also limited at their current levels with help of OPC UA converters and Tunneller. Some of the products written on very old technologies that required outdated infrastructure to function were also replaced. The targeted security level was achieved by restricting the OPC UA TCP\IP ports to open only in the DMZ layer. The OPC UA Data Bridge used in the DMZ Layer allowed:

1. Consumption of OPC UA TCP\IP data without any data loss or distortion.

2. Ready to use connectivity to plant historians and Azure cloud.

3. Only upward movement of data with all the firewalls forts configured to be outbound.

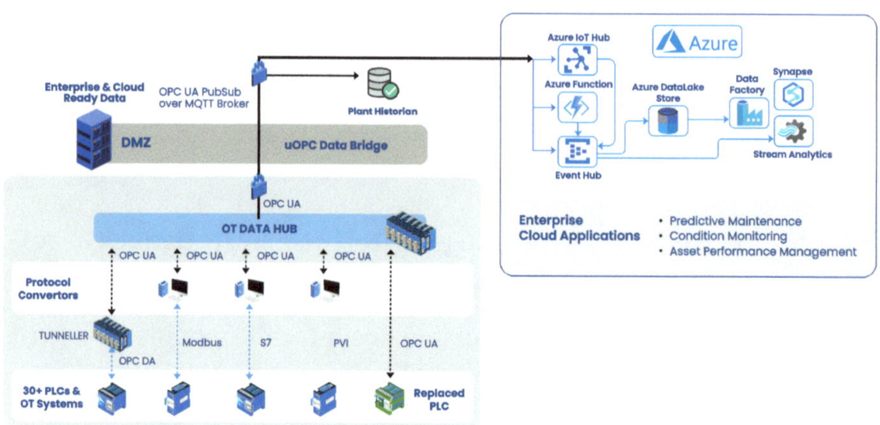

Illustration: OPC UA-Based Harmonized Cloud-Data Movement Setup

This redesign helped standardize the OT setup with only on 4-5 standard/supported protocols and OPC UA TCP/IP as the pre-dominant protocol at the OT Network level. Fewer protocols meant Firewalls were more secure at different levels.

Removal of legacy protocols reduced maintenance costs. Better devices and converters reduced breakdowns and loss of data by about 60%. The data was not distorted, and samples losses were much lesser while moving data from L1 L2 level to DMZ level.

Case Study 3:

A COMMUNICATION INTERFACE USES OPC UA FOR COMPREHENSIVE INFORMATION MODELLING

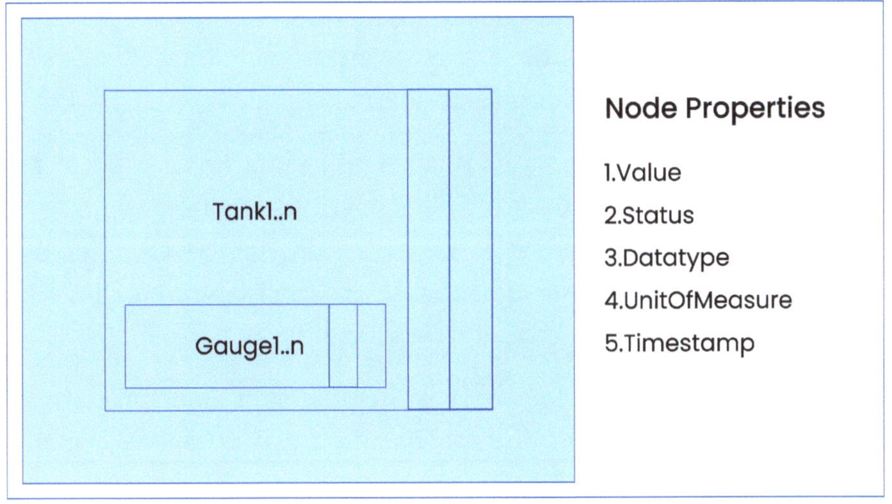

Illustration: Tank Gauge User Requirement

An industrial automation device manufacturer wanted to upgrade their communication interface. Their SCADA software embedded box was used by tank farm operators to monitor and measure real-world tanks and gauges. However, their device used a very old field protocol – Modbus – at the lower levels and was unable to support the new market needs and was prone to cyber risks. Also, each box could only monitor a limited number of tanks and it was difficult to collate this data across facilities.

Each SCADA software was connected to multiple tanks, and each tank had multiple gauges in the real scenario. The tank farm operators needed an information model to visualize the same hierarchy. The ask was a digital twin of operations where information moves all the way from their real-world assets to the cloud and to user dashboards. To support this requirement the client wanted to develop an OPC UA Server in SCADA software as a secure communication mechanism between the host systems and SCADA software in addition to the existing Modbus TCP/IP communication mechanism. The intention was to expose the data to an OPC UA client-based system and provide a mechanism to execute commands from a remote system.

The OPC UA server was built to enhance efficiency, improve security, and amplify edge computing capabilities. An Information Model was also built to map the customer's Assets to OPC UA types to get the contextual data. This information model was built based on OPC PA-DIM information Model companion specification.

Before shortlisting PA-DIM as base companion specification, a comparison of a few companion specifications carried out to help the client understand the best fit for their needs. From the detailed study and deliberation, the PA-DIM information model was zeroed in.

The built-in information model reflected the data from the physical objects as closely as possible in real-time. The relationships between assets starting with the base object type were defined. Followed by modelling the data structure, various components were created and extensively tested to ensure that everything worked correctly as expected. The transition to a cross-platform setup was also necessary, involving the use of a Linux-based server that couldn't accommodate all the profiles and had limited memory.

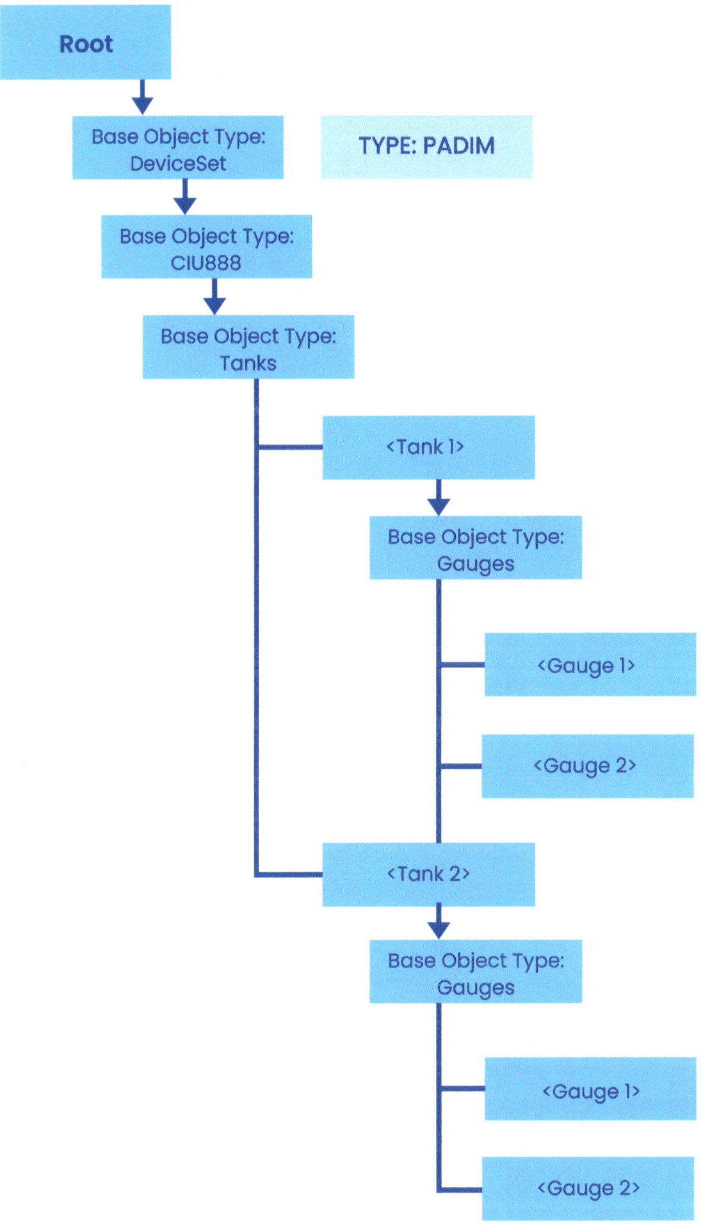

Illustration: PA-DIM-Based Tank Gauge Hierarchy

The new system was able to handle more data without losing context or any information, it was more secure, and it gave the users better control over their assets.

A similar project was delivered a similar project for an engineering company's tank gauging system that operated on tank master software and wanted to build an information model.

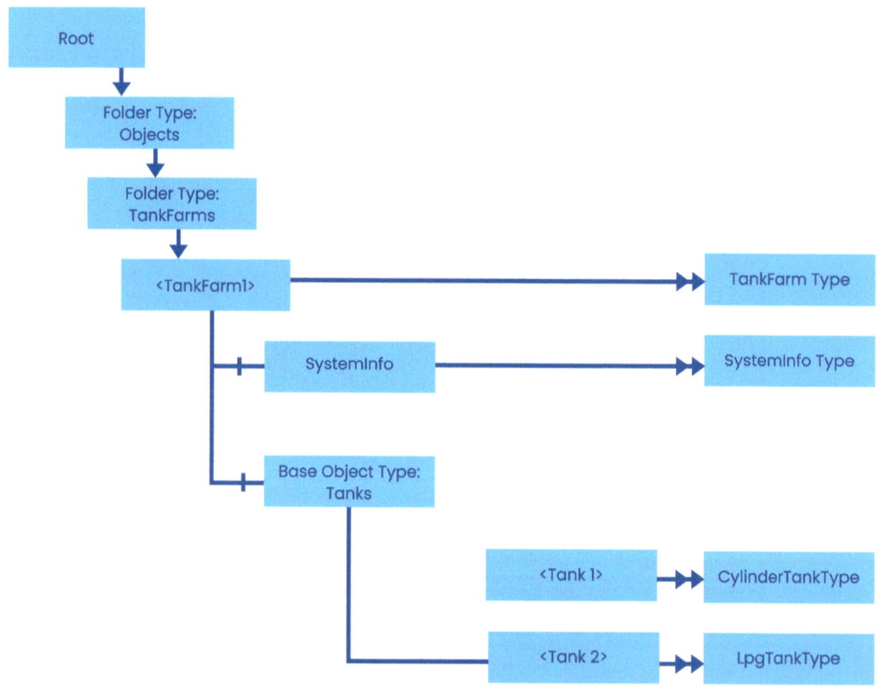

Illustration: PA-DIM-Based Tank Farm Hierarchy

From numerous such deployments, OPC UA has created and delivered tremendous values for our clients across industries. In the next sections, we will explore some key elements of this framework that help deliver these outcomes and make information modelling in such detail possible.

Case Study 4:

BUILDING A DYNAMIC INFORMATION MODEL FROM OPC CLASSIC ADDRESS SPACE

One of the leading global OEMs wanted to expose the complete information of parameters through the OPC Classic DA Server. However, OPC Classic did not support the dynamic information model.

To overcome this, a customized solution was proposed using Utthunga's uOPC Aggregation server – a multipurpose OPC Junction - to overcome the challenges faced by the legacy component running the legacy OPC D.A Classic Server which has no support of any kind of Information Modelling. The uOPC Aggregation server acts as a OPC UA wrapper and converts the OPC classic server data outputs into OPC UA outputs. OPC UA can then organize multiple information models that are optimized for a class of client.

An OPC UA Client was installed in a centralized location and established a secure connection with all uOPC Aggregation servers installed in different site locations. This enabled conversion from OPC DA to UA Server with ability to build Dynamic Information Model based on the Address Space exposed by the end Server.

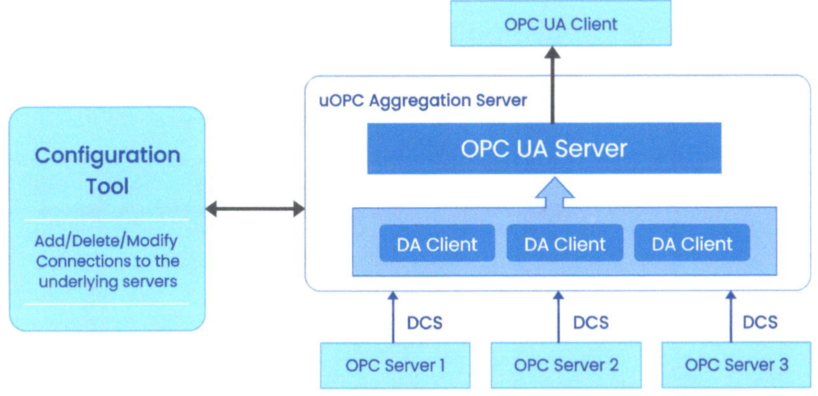

Illustration: OPC UA Information Model-Enabled OPC UA Aggegration Server

Case Study 5:

A GLOBAL LEADER IN COSMETICS IMPROVES OPERATIONS PERFORMANCE WITH OPC UA

A cosmetic market leader with a diverse brand portfolio, embarked on a digital transformation journey aimed at enhancing operations and meeting business objectives. Their goals included addressing daily operational challenges faced by their staff, from operators to managers, bolstering cybersecurity, and building a 4.0 architecture. They sought to create a bridge between equipment sensors and operational platforms, while also aggregating consumer data into a digital tool. The end-goal? A comprehensive dashboard for operators and an analytics tool for the performance team.

OPC UA was chosen as the foundation of their digital architecture. This enabled the cosmetic giant to clearly define roles and responsibilities between the company and its supplier ecosystem. It also helped standardize the OT architecture of their digital tools.

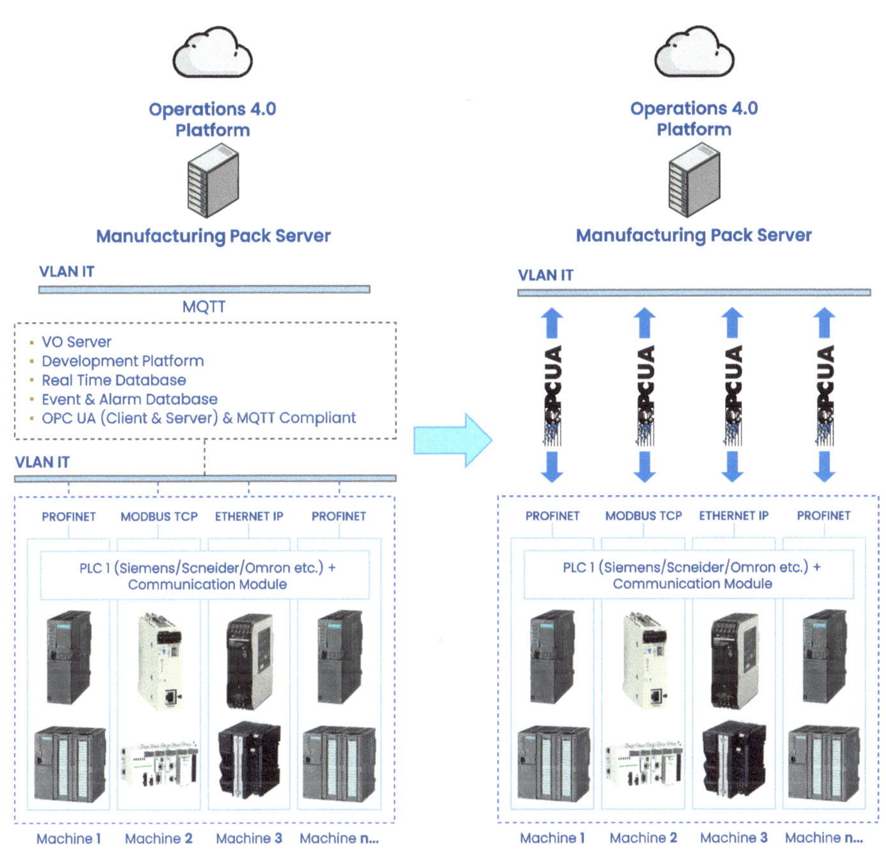

Reference: OPC Foundation

On the ground, in areas such as filling and packing, they faced the challenge of having multiple machines from varied suppliers, all operating on different hardware solutions like PLC and HMI. With OPC UA, the company developed a "connected line tool" that aggregated data from each machine, offering operators and technicians real-time insights. This tool eased the monitoring of individual machines, reducing stress for operators and boosting efficiency.

Another significant tool introduced was the "chrono analysis" application. This powerful tool focuses on micro-strategies of machine management, offering visibility into real-time machine performances and possible disruptions during the production phases.

Additionally, they developed a quality process management tool for premium packing lines called "track and trace." This was piloted on lipstick lines to monitor potential variations during production, aiding quality assurance teams in swiftly identifying and rectifying issues.

Thanks to the OPC UA architecture and their dedication to digital transformation, the company has implemented over 4030 different connected machines globally. The benefits have been manifold – from team workload reduction to increased operational efficiency (OEE). In some cases, the OEE target improved by 2 to 10%, depending on the line complexity.

The company's ambition doesn't stop there. Future plans include expanding the 4.0 tools to address energy consumption monitoring, enhance data management between machine data and manufacturing execution systems, and improve the environmental impacts of their operations.

This digital transformation, anchored by OPC UA, showcases the power of leveraging technology to drive operational excellence in a global manufacturing environment. This journey not only optimized their current operations but set the stage for even greater advancements in the future.

Case Study 6:

A MULTINATIONAL AUTOMOBILE MANUFACTURER GATHERS LARGE-VOLUME INDUSTRIAL DATA

A French multinational automobile manufacturer recognized a challenge in their factories: each object, from screwdrivers to packaging, had its own unique data and interface. Names and interfaces varied, causing confusion. To standardize and connect these items, they pioneered an industrial metaverse, a digital twin of their supply chain. This simulation offers real-time insights into operations, allowing timely interventions, increasing resilience, and cutting costs by 20-30%. Benefits include reduced order-to-delivery times, optimized energy use aiding in carbon footprint reduction, and enhanced employee perspective.

However, they faced daunting transformation challenges in scaling industry 4.0, particularly in capturing the clean and intelligent data. Legacy systems, variety of OT systems, protocols, and solutions to manage shopfloor operations along with complex IS integrations made it difficult to get contextual data. Moreover, the company lacked data standardization both at machine and process level. In addition, connecting OT and IT in both directions was difficult. Different tools, networks, and cultures compounded the problem.

They started with a conditional and predictive maintenance project to structure and standardize data choosing OPC UA for data modelling. They created a specific IOT gateway for existing equipment to transform existing data to OPC UA data model. The OPC UA data model was easy to use for people in the plant.

After the success of this program, they launched a group initiative around industrial data management across all organizational levels which was also a success. At the facility level the created an OPC UA unified data collector (uDC) that collected data from proprietary protocols and converted it to OPC UA structured data model and unique vocabulary.

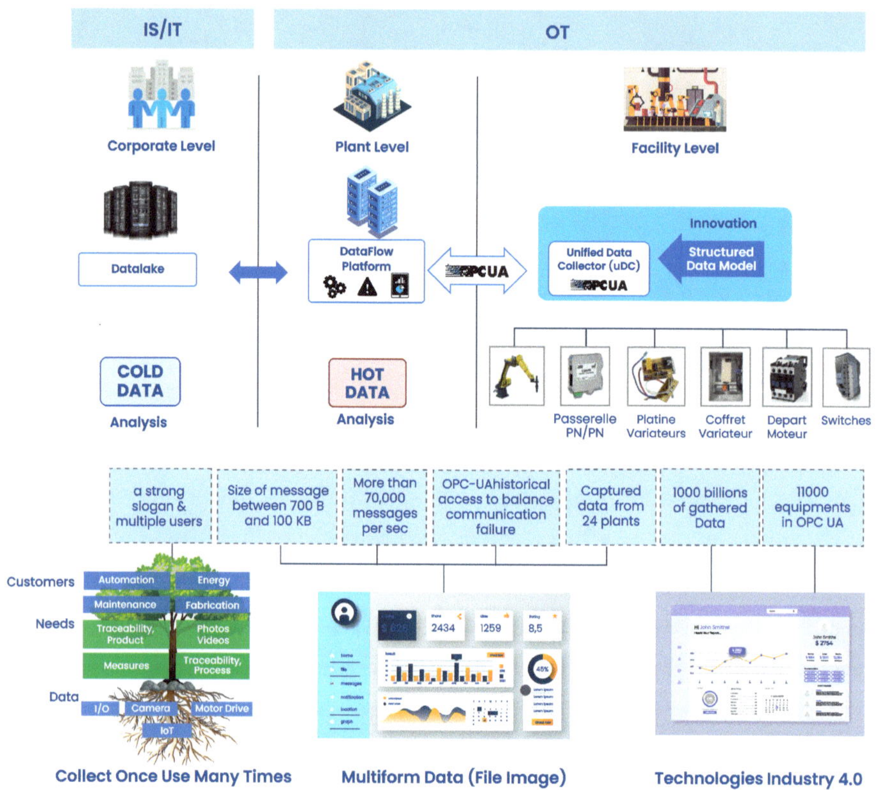

Reference: Illustration #6

At the plant level in IS/IT layer they created a possibility to monitor hot data like real time energy consumption. At the corporate level, they created the capability to analyze cold data to understand the need for quality management, maintenance, etc.

They now have thousands of equipment working on OPC UA model. Also, several billion points of multiform data is collected from 24 plants. They also use OPC UA historical access to balance communication failure.

• • •

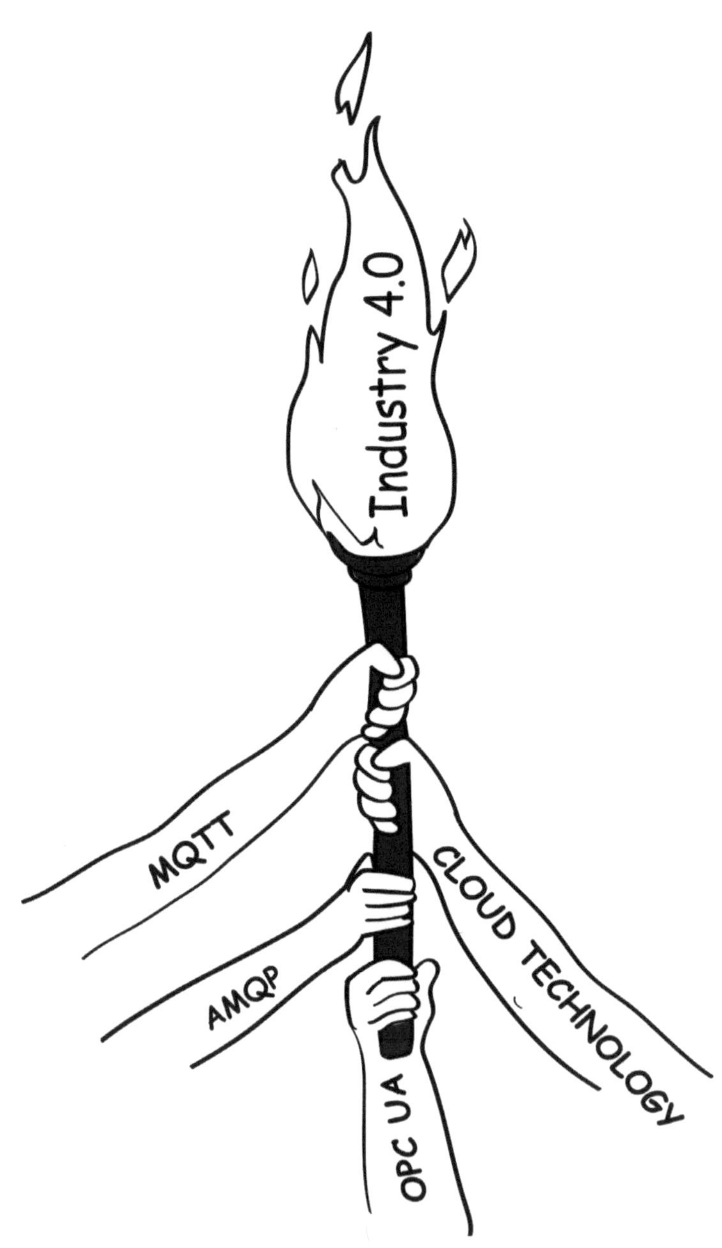
Industry 4.0
MQTT
AMQP
OPC UA
CLOUD TECHNOLOGY

OPC UA Myth Busting #5

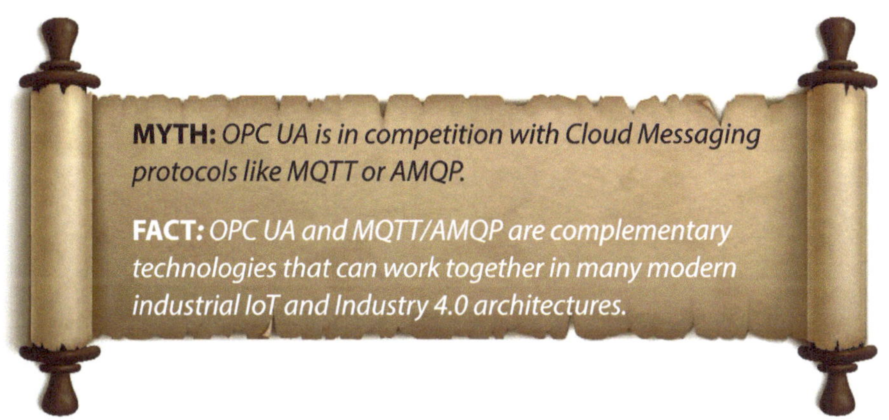

The myth that OPC UA is in competition with popular messaging protocols for Cloud such as MQTT and AMQP likely originated from a misunderstanding of their respective roles and the contexts in which they are typically used. While multiple articles on the Internet support this false claim, the truth is that OPC UA's Pub-Sub mechanism supports these messaging protocols for network communication.

Both OPC UA and MQTT/AMQP are involved in the transmission of data, but in different contexts and for different purposes. This distinction can blur, especially as the boundaries between traditional industrial automation systems and the emerging IoT and cloud technologies become increasingly intertwined.

However, saying OPC UA competes with MQTT or AMQP is an oversimplification. OPC UA and MQTT/AMQP are complementary technologies that can and do work together in many modern industrial IoT and Industry 4.0 architectures.

OPC UA's strengths lie in its representation of complex data and its built-in information modeling capabilities. It provides a sophisticated method for devices to describe and share their capabilities in a standardized way. This makes OPC UA great for understanding the context and meaning of data. In contrast, MQTT and AMQP are messaging protocols that excel at reliably and efficiently transmitting messages, particularly in distributed or cloud-based systems. These technologies are often used together, with MQTT or AMQP providing a lightweight, efficient transport mechanism for OPC UA data.

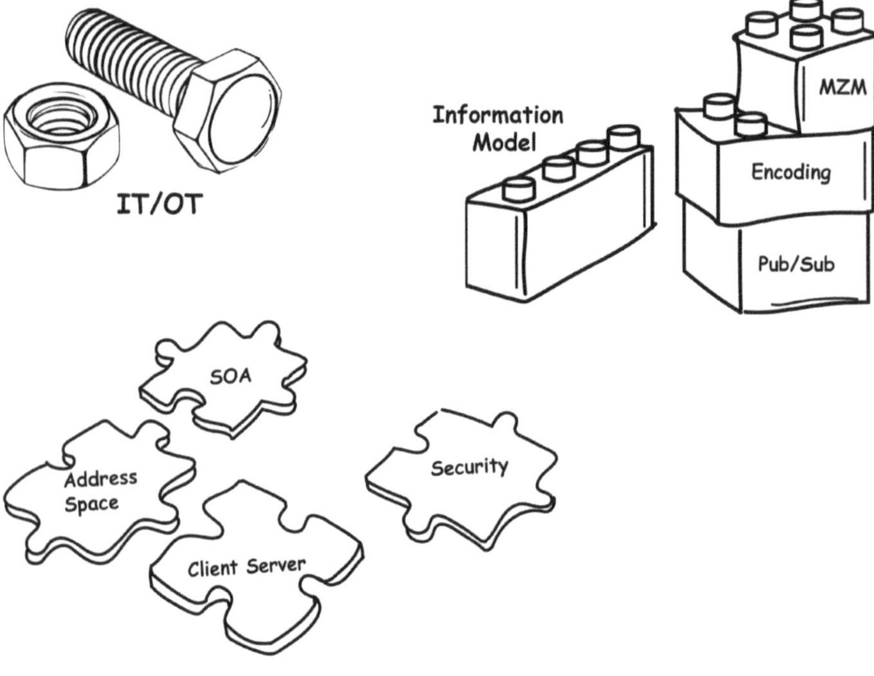

IT/OT
Information Model
MZM
Encoding
Pub/Sub
SOA
Address Space
Client Server
Security

The Nuts & Bolts of OPC UA

KEY DESIGN ELEMENTS OF OPC UA

To meet the growing interoperability needs, OPC UA was designed to be platform-independent with a firewall- and IIOT-friendly service-oriented architecture (SOA) and a comprehensive information model.

Platform Independence to Support Interoperability Across Systems:

Companies operate on a wide array of hardware platforms and operating systems. Platform independence ensures OPC UA Server compatibility on those platforms and devices.

The OPC UA specification is technology agnostic and can be implemented in any technology* necessary to support specific platforms** or in non-OS operating systems. This versatility enabled OPC UA to be embedded in switches, sensors, gateways, embedded controllers, edge devices, etc.

Software Programing Friendly:

*ANSI C, C++, Java, .Net, Python, etc.

**Linux, Android, iOS, Windows etc.

Firewall-friendly SOA for IT/OT Convergence:

OPC UA overcomes the limitation of OPC Classic and allows secure exchange of information over firewalls between OT and IT networks. OPC UA supports a Service Oriented Architecture and uses well defined OPC UA service interfaces to facilitate seamless communication between OPC Server & Client components over various networks.

The OPC UA specification acts as a bridge between tightly coupled and loosely coupled systems to move data from the factory floor to the enterprise cloud. It is designed such that OPC UA data that is being exposed or transferred is independent of the underlying transport mechanism being used. The transport mechanism can be configured during the OPC UA application deployment stage by the field engineer.

The open nature of the OPC UA specification helps make upgrades and maintenance easier. The service will not expose how it executes its functionality, just what it does. It also ensures that the information model and transport mechanisms are completely independent of each other. This helps in enhancing the standard as per future needs of the market and technology advancement.

Comprehensive Information Model That Can be Extended for Unique Needs:

In real world use cases, there will be correlation between some of the real-time data being read from the field device and the alarms/ events being generated. For example, if a 'Flow Out of Range' alarm is generated in a field instrument, it is because the 'Outflow of the Pump' parameter value has changed. The OPC UA information model exposes these correlations as unified data access. In addition, it exposes data in a complex information model that supports the correlation of parameters and multiple hierarchies. And finally, this information model can be extended based on the specific domain and manufacturer needs.

IIOT-friendly Architecture that Works on the Most Demanding Production Systems:

The OPC UA standard has been designed to suit the current needs of IIoT. Apart from the Client/Server architecture, the latest specification of OPC UA supports OPC Pub/Sub mechanism and deterministic communication via TSN. These developments have helped create an open architecture that works even for complex and most demanding production systems.

OPC UA COMMUNICATION MECHANISM

OPC UA has a defined set of sophisticated communication mechanisms to transfer data and information securely between different devices and systems. In the coming pages we will explore some of these elements of this mechanism in detail.

OPC UA Messages	OPC UA Endpoints	OPC UA Sessions
Messages are exchanged between client and server via endpoints and are used to transfer data and information between different devices and systems.	Endpoints are secure communication channels with a unique address that allow clients and servers to exchange messages over transport protocols.	Sessions are logical connections established between a client and a server to exchange messages and maintain state information.

OPC UA Security	OPC UA Address Space	OPC UA Services
Security mechanisms protect the communication between client and server via authentication, authorization, encryption, digital certificates, and key exchange protocols.	Its a hierarchical structure that represents the information model of a system. Clients use the Address Space to retrieve information about the system and its components.	Services are a set of predefined operations such as Read, Write, Browse, Subscribe, Call, etc. that can be used to interact with the system.
OPC UA Discovery	**OPC UA Encoding**	**OPC UA Transport Mechanism**
Discovery mechanisms facilitate the identification of and communication with servers on a network. OPC UA discovery includes Multicast DNS, and LDAP.	Encoding is the process of converting OPC UA data into a format that can be transmitted over a communication channel. OPC UA supports several encoding types, including binary encoding, XML encoding, and JSON encoding.	Transport Mechanisms such as TCP and MQTT are used to transmit OPC UA messages between client and server applications over a communication channel, such as a network or a serial connection.

OPC UA Audit Trail
Audit Trail records all user actions and events related to the OPC UA system in a secure and tamper-proof manner. This provides a complete history of system activities.

DEEP DIVE INTO OPC UA ADDRESS SPACE AND INFORMATION MODEL

Information modelling is the heart of OPC UA and differentiates it from the rest of the communication protocols. OPC UA is not just a transport protocol, but a framework defined based on an open specification that allows OPC UA Servers to describe the underlying system in a more meaningful way. It allows us to exchange not only the data, but also semantic information associated with the data with a comprehensive information model and address space.

The information model defines the structure and organization of the data being exchanged, while the address space provides a hierarchical structure for accessing and organizing that data. Together, these concepts provide a standardized way of exchanging data between different industrial automation systems regardless of the underlying hardware or software platform.

The **semantic address space model** defines how information is structured and exposed by OPC UA servers enabling a consistent and meaningful representation of information across different devices and systems.

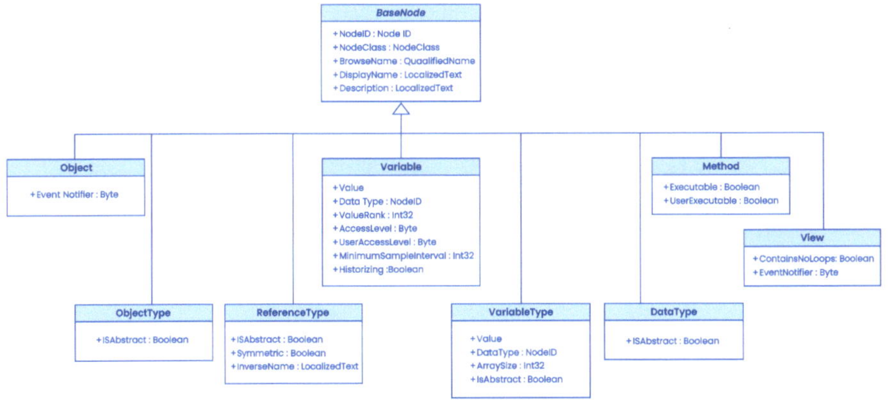

Illustration: OPC UA Information Model Base Classes

The address space is a hierarchical structure modelled as a set of Nodes accessible by clients using OPC UA Services. Nodes in the address space are used to represent real objects, their definitions, and their references to each other[6]. Depending on their purpose, Nodes can be of different node classes - Object, Variable, Method, ObjectType, VariableType, ReferenceType, DataType, or View. There are also nodes representing instances and others representing types, etc.

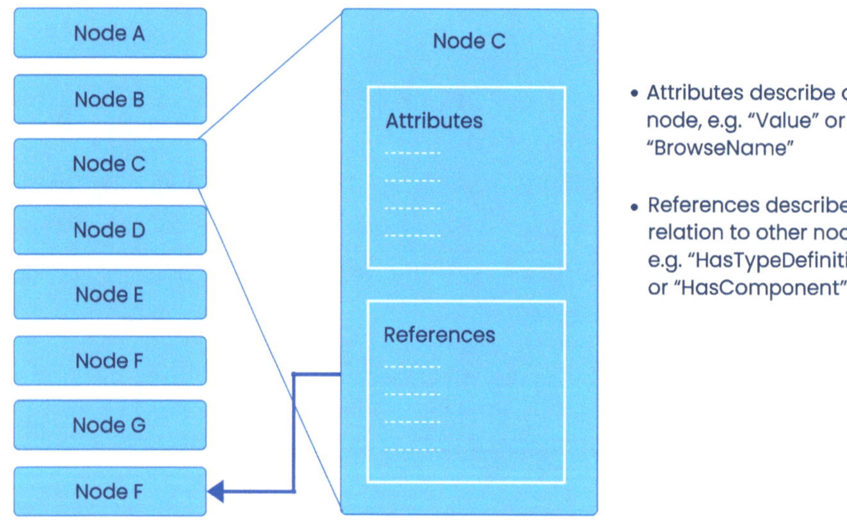

Reference: Illustration #7

The OPC UA address space allows discovery, browsing, reading, writing, calling, and monitoring of nodes by OPC UA clients. As it follows the full mesh network model, a single node in OPC UA address space can be accessed via multiple node paths.

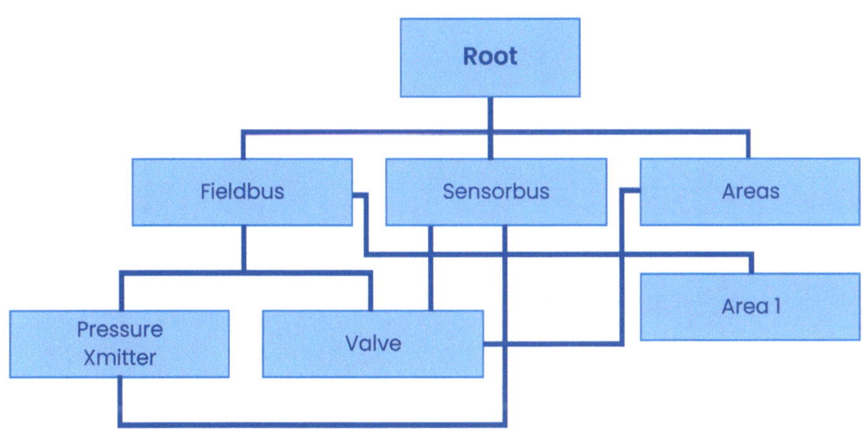

- Full Mesh - Network Model

- Unlimited Named/Typed Relationships

- "Views" are used to present hierarchies

Reference: Illustration #8

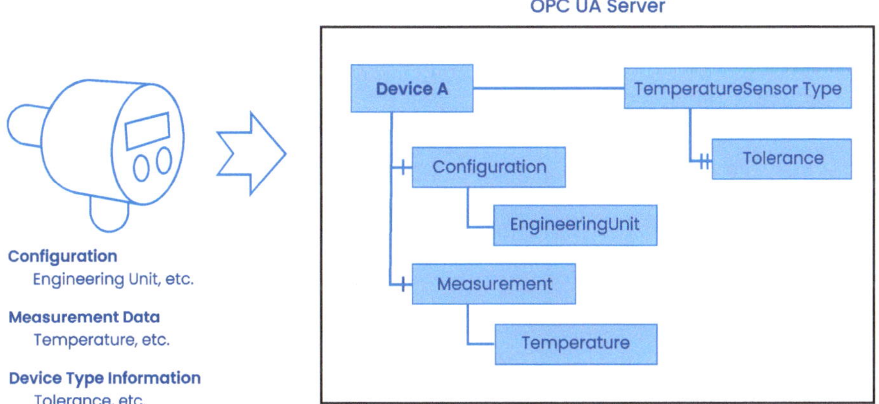

Reference: Illustration #9

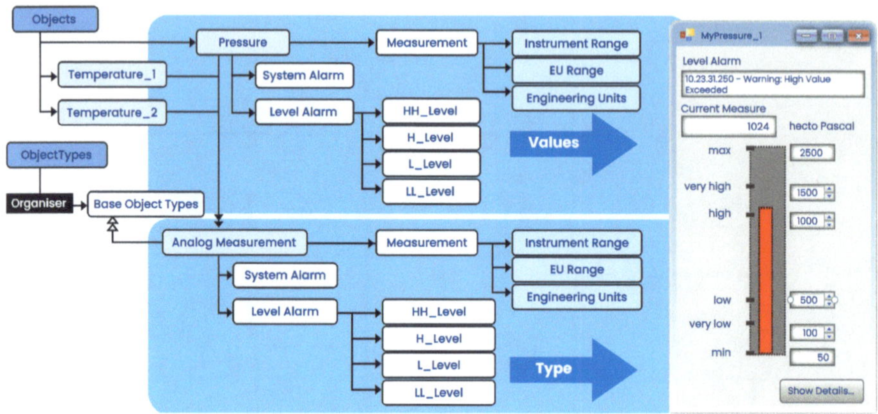

Illustration: OPC UA Types and Objects

The OPC UA **Information Model** is based on the principles of object-oriented programming. Each variable, event, and method in the underlying process is an object that is interconnected by references. Using the object-oriented model makes it easy to extend and customize the information model to meet the specific needs of a particular application or industry. All you need to do is define new Object Types and Variable Types. In the figure above, a real automation instrument – a temperature sensor - with its different properties has been modeled in OPC UA Information Model.

Once the base information model was defined, OPC UA also defined a set of e extended information models for common domains and applications, such as data access, alarms and events, and historical data access.

- **OPC Data Access (OPC DA)** is perhaps the most important OPC interface and used in 99% of the products using OPC technology. OPC DA describes how to access current process data within an OPC UA system and defines the structure and behavior of data such as how data is read, written, and monitored. Data Access is used to exchange real-time data such as sensor readings or machine status updates from PLCs, DCSs, and other control devices to HMIs and other display clients.

- **OPC Alarms and Events (OPC AE)** allows the client to receive event and alarm **notifications** providing a flexible interface for transmitting process alarms and events from different sources. OPC AE helps companies define standards to represent alarm states, acknowledge alarms, and define the conditions under which alarms are triggered.

- **OPC Historical Data Access (OPC HDA)** provides easy and systematic way to access, retrieve, and query data stored in historical archives. OPC HDA can read historical data in three ways – raw data, one or more variables for a specific timestamp, and aggregated values for a specific time domain for one or more variables[7]. Other than reading the data, OPC HDA also enables inserting, replacing, and deleting data in the history database.

Note: There are many more profiles defined in OPC UA specification but for the simplicity of this book the focus would be on the above three profiles.

Enabling Decisions in a Steel Plant with OPC UA Information Model

OPC UA information model and address space provides invaluable insights creating an accurate and detailed view of a plant or facility. These real time insights can be used to make real-time decisions that can improve production efficiency. For instance, the information model for a steel plant's production processes and equipment can enable informed decisions about plant operations and maintenance.

Let's consider the production of steel sheets in a steel plant. This process has several steps such as melting the raw material, casting the molten steel into slabs, and rolling the slabs into sheets. These steps require several equipment such as furnaces, casting machines, and rolling mills.

With the OPC UA information model, the company can create a detailed representation of this process and equipment with machines, sensors, and variables arranged hierarchically. For example, a furnace object may contain variables such as temperature and pressure, while a rolling mill object may contain variables such as speed and torque. The OPC UA address space provides a unique identifier for each object in the information model and allows OPC UA clients to access and control these objects.

For example, an OPC UA client can read the temperature or pressure variables of a furnace object to figure out whether it is operating at optimal conditions. If the conditions are not what they should be – for instance temperatures running low – it can alert the maintenance manager who can schedule repairs before a production delay happens. Similarly, the data from speed and torque variables in the rolling mill object can help identify whether the rolling mill is working as it should. If not, production engineers can be alerted to adjust the mill and ensure that the steel sheets are produced to the correct specifications.

THE OPC UA CLIENT/SERVER ARCHITECTURE

The OPC UA client/server architecture is a distributed system where OPC UA clients and servers communicate with each other based on a set of standardized messages and services. Here is how it works:

- The OPC UA client sends requests to the server, and the server responds with the requested data or services. This communication can be initiated by either the client or the server, depending on the specific use case.

- OPC UA Server components fetch data from real world objects. For example, motor speed, positioner setpoint etc. and build the OPC UA address space.

- OPC UA Server SDK (Software Development Kit) provides the API to expose this OPC UA address space via OPC UA Services. Using these APIs, any OPC UA client can query and browse the OPC UA address space and perform read/write or subscription operations.

- Single OPC UA Server applications can integrate real-time data, history and events associated with the data.

- In addition, OPC UA Server can be configured to limit the number of concurrent OPC UA client sessions connecting to it.

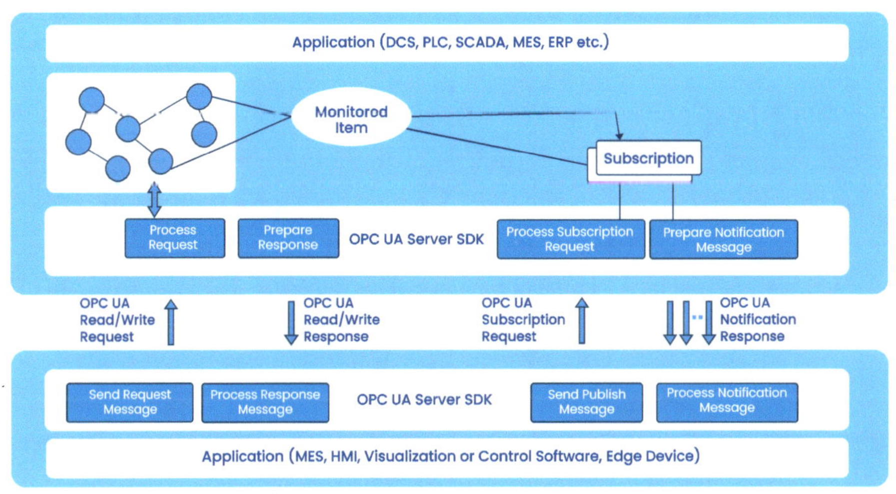

Reference: Illustration #10

Monitoring and Controlling an Automotive Assembly Line with OPC UA Client/Server Architecture

In the automotive industry, the assembly line is a critical part of the production process. Various machines, robots, and sensors work together on the assembly line to put together the desired vehicle. Real time monitoring and control of the assembly line enable prompt action and prevent unexpected downtime or issues that could impact yield or quality. Let's see how the OPC UA client/server interaction can enable this.

Once the manufacturer deploys OPC technology, the assembly line is connected to an OPC UA server. This server collects data from various sensors and machines and sends it to an OPC UA client that in this case would be a supervisory control system. Once connected, this supervisory control system or the client receives real-time data from the assembly line to monitor what is going on. For example, what is the temperature and pressure in the paint booth, or the speed and torque of the robotic arms assembling parts.

The OPC UA client and server can also work together to control the assembly line. For example, if the OPC UA client detects some anomaly, it can send a command to the OPC UA server. These commands could be related to starting or stopping the assembly line, adjusting the speed of the conveyor belt, activating a particular machine, or something else. The OPC UA server can then send these commands to the appropriate machines or sensors to perform the desired action.

Server-to-Server Interaction

An OPC UA Server can communicate with another by acting as both a Client and a Server. For example, an Edge Gateway uses the OPC UA Server interface at the downlink layer to gather information from the network through its OPC UA client interface, and then relays the processed data to an uplink cloud solution using its OPC UA Server interface.

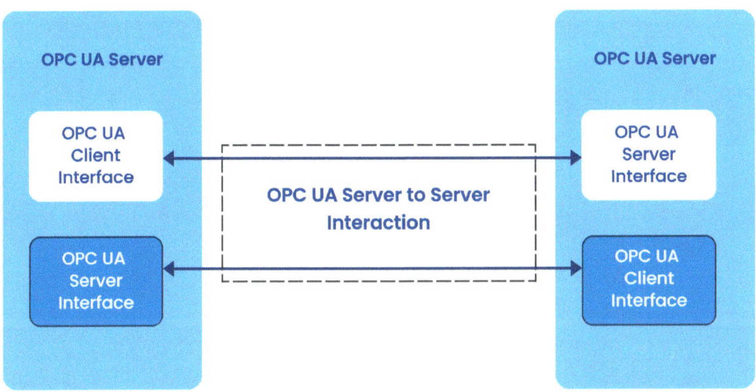

Reference: Illustration #11

Chained Servers can connect to multiple OPC UA servers across locations and networks in a chain, aggregating the data and presenting it as a single, unified data source to an OPC UA client.

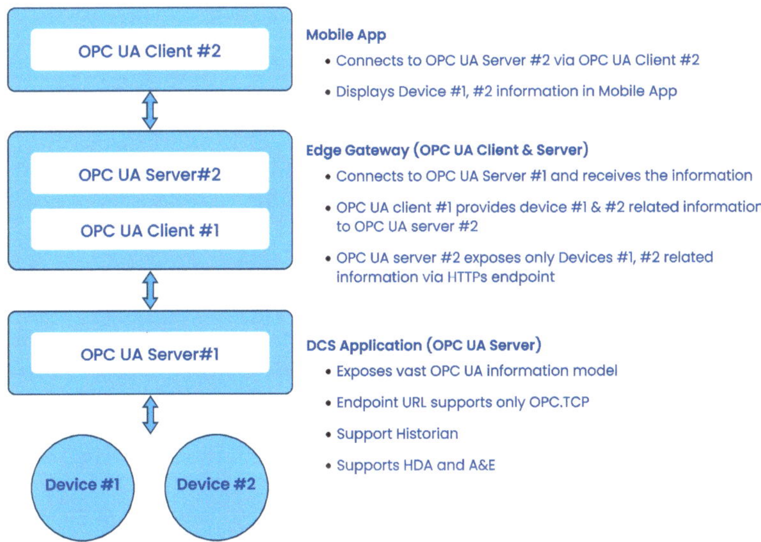

Illustration: Chained OPC UA Server Architecture

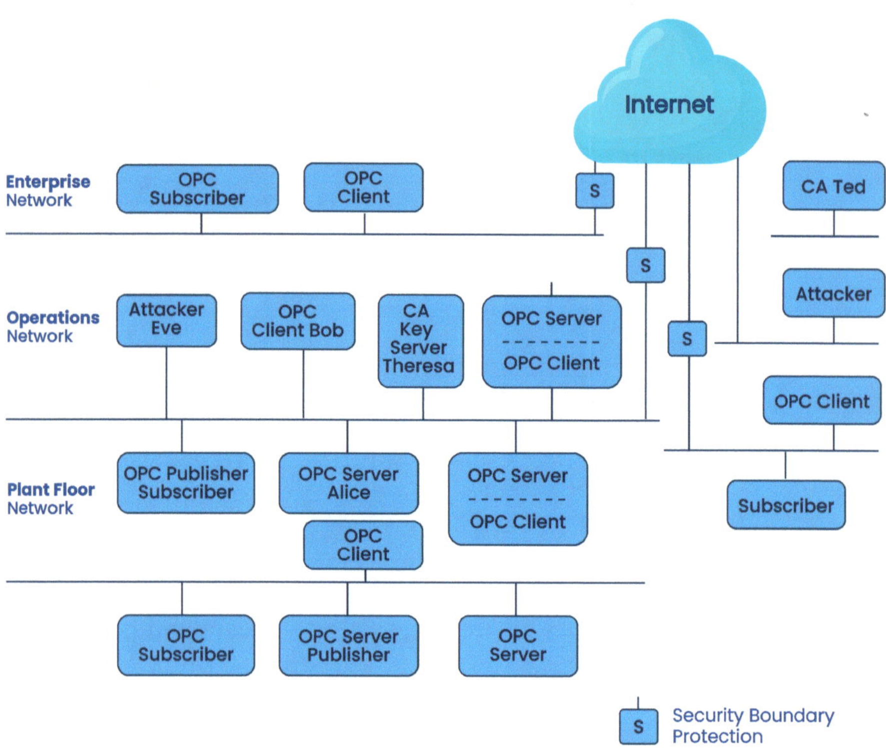

Reference: Illustration #12

Aggregation Servers also collect, consolidate, and analyze data from multiple OPC UA servers but in the same network and geographic location.

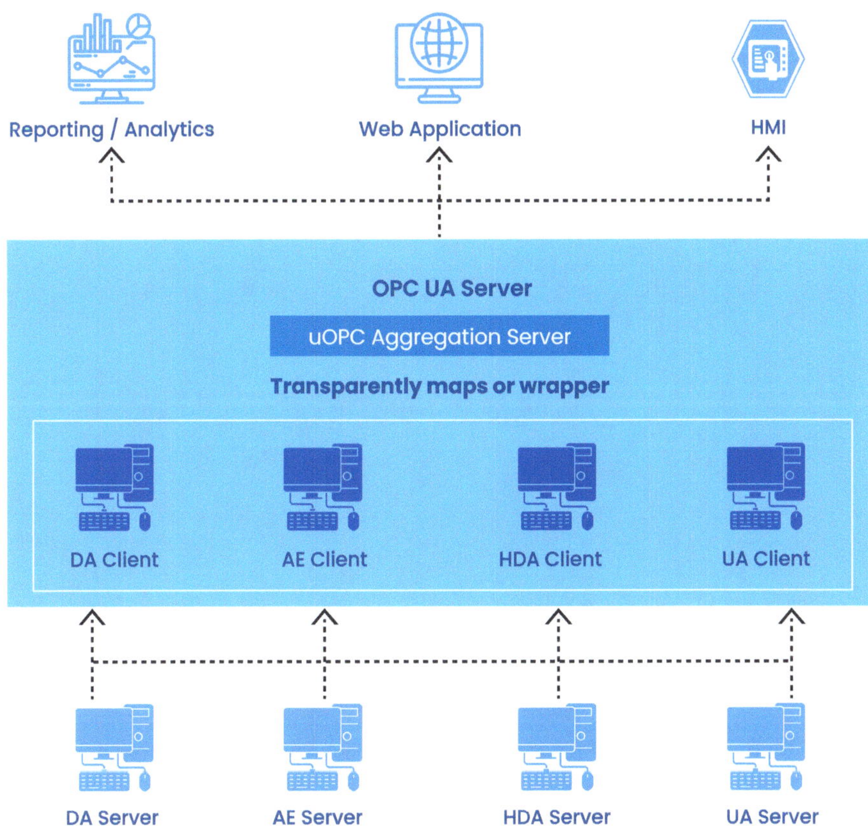

Illustration: uOPC Agreegation Server

Redundancy Servers ensure that data and services remain available in the event of a server failure. They have multiple servers in a primary-secondary configuration to enable high availability and fault tolerance. If the primary server fails, the secondary server takes over and continues to provide data and services to the OPC UA clients.

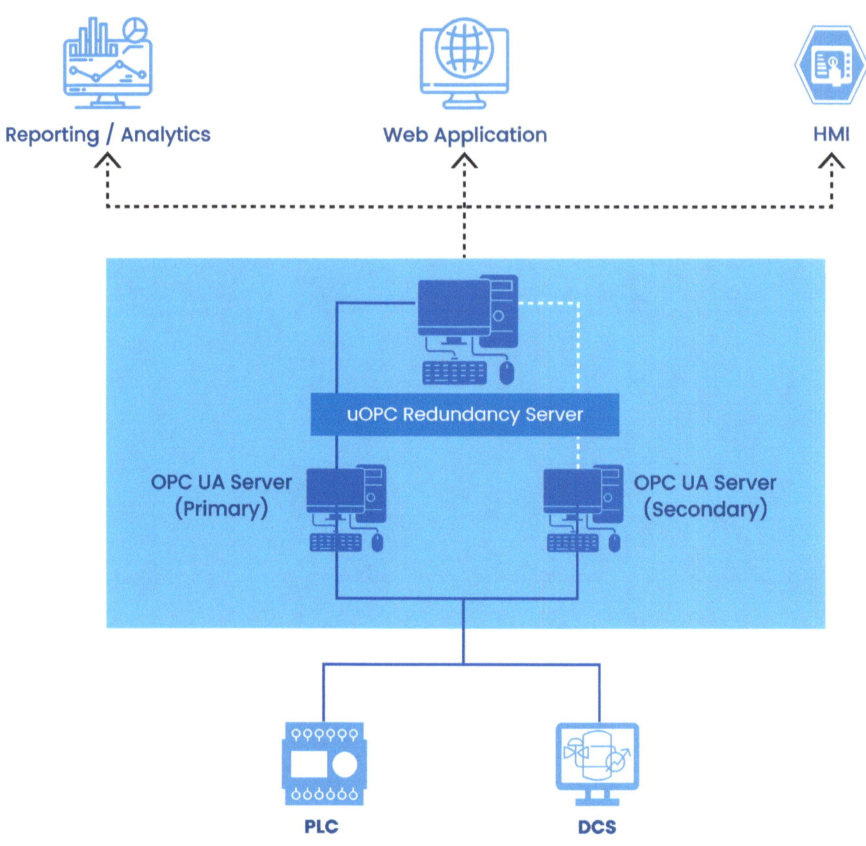

Illustration: uOPC Redundnecy Server

UNDERSTANDING TRANSPORT LAYERS AND DISCOVERY SERVICES

OPC UA supports multiple transport layers and discovery services that enable secure and reliable communication between different industrial automation systems. The support for multiple transport layers (including TCP), allows OPC UA to be used in a wide range of systems regardless of the underlying communication protocols. The discovery services enable OPC UA clients and servers to find each other on a network. Once an OPC UA client fetches the information on the interested OPC UA server endpoints, it will be reused for further connection requests to avoid multiple queries to the discovery server.

There are three main discovery services in OPC UA: Local Discovery Server, Multicast Discovery, and Global Discovery.

- Local Discovery Service: In this mode the OPC UA Server and Clients are in the same machine. Single Discovery Server can support multiple servers running on the same machine from multiple vendors.

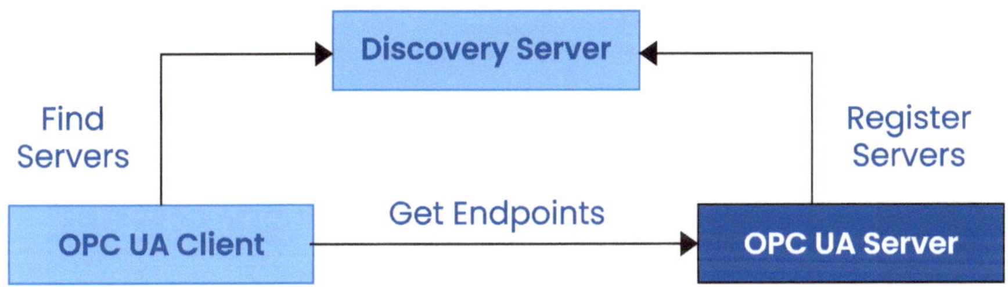

Reference: Illustration #13

- **Local Discovery Server with Multicast Extension:** Using the LSD-ME mechanism, OPC UA clients can discover the OPC UA Server located in different machines in the same network. This feature does not require the central infrastructure; however this will work only in the local subnet.

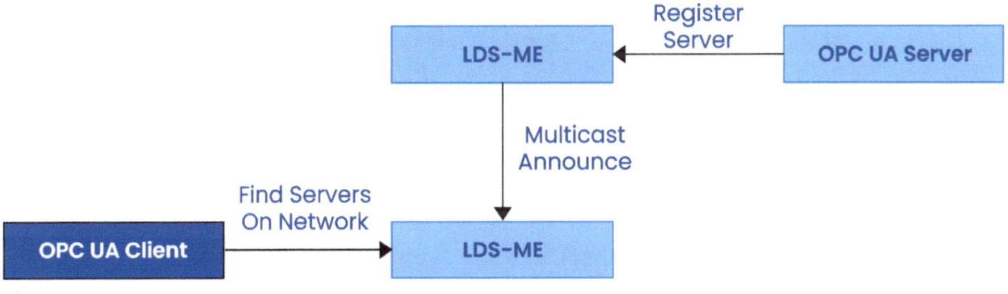

Reference: Illustration #14

- **Global Discovery Server:** GDS allows the OPC UA Client to discover any OPC UA Server globally. OPC UA Server can be distributed at several geographical locations in different networks. While querying the OPC UA Server by the OPC UA client, it can provide the filters to optimize the server list returned by the server.

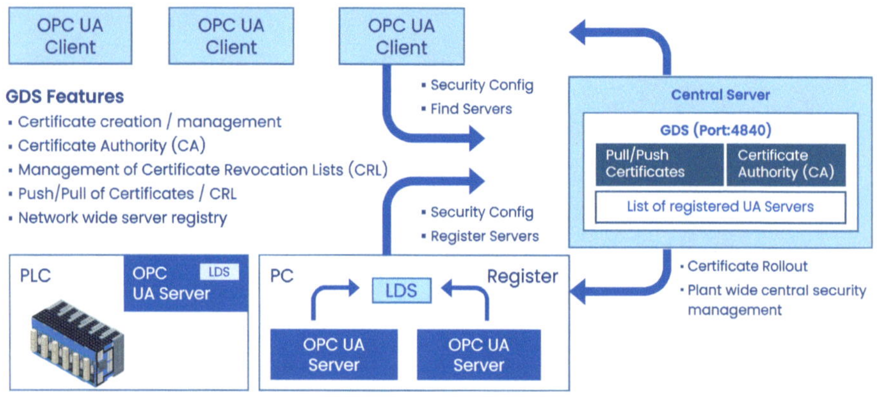

Illustration: Global Discovery Server (GDS)

The advantage of using OPC GDS Server is that it can also act as a Central Certificate trusting authority and let other OPC UA Clients and Servers register their certificate into their trust list and get the certificate from their trust list.

OPC UA SUBSCRIPTION VS OPC PUB/SUB

An important aspect of OPC UA is its support for data subscriptions and publishing. OPC UA subscriptions and OPC Pub/Sub are two different mechanisms for receiving data updates from an OPC UA server.

OPC UA Subscription can be set up by an OPC UA client to receive regular data changes from an OPC UA server. For example, an OPC UA client can set up a subscription to receive temperature readings from a sensor every second. In simpler terms, OPC UA subscriptions are like subscribing to a magazine where you get regular updates from the publisher.

However, in this mode, the OPC client application must constantly maintain the session with OPC UA server. While that is okay for medium scale subscriptions, it quickly becomes unmanageable as the number of OPC UA subscriptions increases. With each additional subscription that needs to be maintained, resources required to maintain these sessions at the server side also increase. This limitation makes OPC UA subscription unsuitable for the demanding needs of IIoT applications.

To overcome this challenge of scale, the OPC Foundation introduced the concept of **OPC Pub/Sub (Publish/Subscribe)** as Part 14 of OPC specification. The OPC Pub/Sub model is a complementary feature on top of the OPC client server model that supports a more efficient way of transmitting data updates.

OPC Pub/Sub is like a news broadcaster. In this model the OPC UA server does not send updates to individual clients directly and instead publishes them to a network, The OPC UA clients subscribe to receive updates from the network. This way, the OPC UA server can send a single update to the network, and all subscribed clients can receive the update simultaneously.

OPC UA PUB/SUB Vs. CLIENT/SERVER

OPC UA Pub/Sub and Client/Server are communication models designed for different types of applications.

OPC UA Client/Server is a traditional request/response-based communication model that is more suited to point-to-point communication between server and client in an OT environment, typically for data access and control applications. In this model the client initiates all communication, and the server only responds to client requests for data.

OPC UA Pub/Sub is a more advanced communication model that can handle high-volume, high-speed, and real-time data streaming. It is ideal for applications with low latency and high scalability requirements where fast response times are critical. In this model, a server periodically publishes data that is sent all subscribed clients.

OPC UA Pub/Sub Vs. Client/Server		
Feature	**Pub/Sub**	**Client/Server**
Data Transfer	Messages	Subscription Basis
Real-time streaming	Yes	Yes
Latency	Low	High
Scalability	Can handle large scale systems	May experience scalability issues

This model is the foundation for OPC UA IIOT use cases. In the OPC classic framework, the OPC client and server architecture was session-oriented. They opened a channel, created a session on top of that, and then used it to pass information. It took a lot of effort and resources to do this, and the system was tightly coupled. To enable IOT use cases, we need loose coupling. The information should be made available to whoever is interested. So, in OPC UA, the server can push the information to a middle agent, or a broker and they will then pass it on to interested subscribers. There is no direct session between the client and server.

OPC UA SECURITY

Security was a fundamental OPC UA design requirement, so it was built into the architecture from ground up. Security mechanisms similar to the W3C Secure Channel concept, were chosen based on the detailed analysis of real-world data security threats and the most effective counter measures against them. OPC UA security addresses key issues like the authentication and auditing of OPC UA clients and servers, message confidentiality, integrity, and availability, and the verifiability of functional profiles.

OPC UA security can be divided into three security levels: User, Application, and Transport. It follows a defense-in-depth strategy, incorporating features like user authentication, encryption, and role-based access control to protect against cyber threats.

- **Encryption:** OPC UA uses advanced encryption algorithms like RSA (for key exchange) and AES (for message encryption) to ensure the confidentiality of the data. This means, even if data is intercepted during transit, it remains unintelligible to unauthorized parties.

- **User Authentication:** OPC UA supports X.509 v3 certificates for user authentication, a widely adopted standard in cybersecurity. Verifying client identities ensures that only authenticated users have access to data.

- **Role-based Access Control:** Giving access rights according to user roles in the organization minimizes the exposure of sensitive data.

Usually, security mechanisms come at a computing resource cost which can adversely impact device performance. The OPC UA standard defines different levels of security (via end points) to enable vendors to implement OPC UA in products with various computing resources. This makes OPC UA scalable. In addition, system administrators can enable or disable such OPC UA server endpoints as required.

Reference: https://opcfoundation.org/

Seven OPC UA Security Objectives:

1. Authentication: OPC UA Clients, Servers, Publishers, Subscribers, users should prove their identities.

2. Authorization: Access to the read/write/execute information should be possible only to systems (OPC UA client, server, publisher, user) that are authorized.

3. Confidentiality: Information is protected

4. Integrity: Receivers receive the same information that the original sender sent

5. Non- Repudiation: Repudiation is the rejection or denial of something as valid or true.

6. Auditability: Actions taken by a system must be recorded in order to provide evidence to stakeholders

7. Availability: System is available to offer the service`

OPC UA: PUTTING IT ALL TOGETHER IN LAYMAN TERMS

The OPC UA workflow involves a client requesting for a service from a server. While the server provides description of the services, the client requests for a specific service and receives the corresponding data from the server. Finally, the server processes the data and delivers the requested service. This workflow can be applied to various industrial automation systems, such as monitoring and controlling manufacturing processes or managing energy systems.

Let's take a hypothetical example to map the OPC UA concept to a real-world scenario. Through this case, let us see how an HMI OPC application displays process values by using the example of a person

(Mr. Paul) wanting to eat at a restaurant. Let's see how the workflows play out.

Paul (an OPC UA Client) is hungry and wants to go to a restaurant (an OPC UA Server) for a meal. So, he opens a Google map (a discovery service) and finds all the nearby restaurants. He now had the option to choose from the many available options. Finally, Paul finds the restaurant he wants to go to, and now he needs some means (a connection to an endpoint) to get there.

Understanding Client/Server Concept in OPC UA:

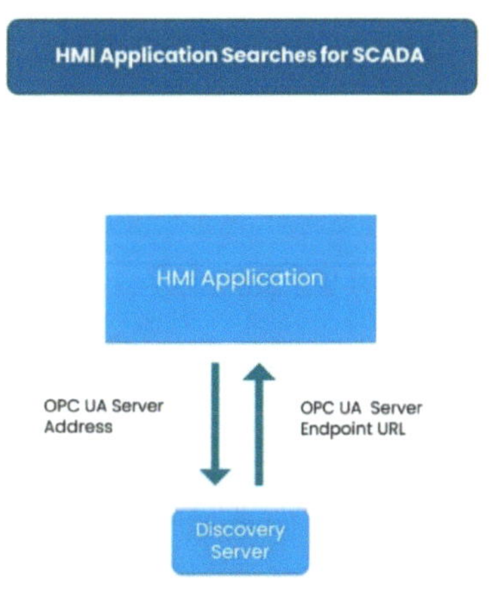

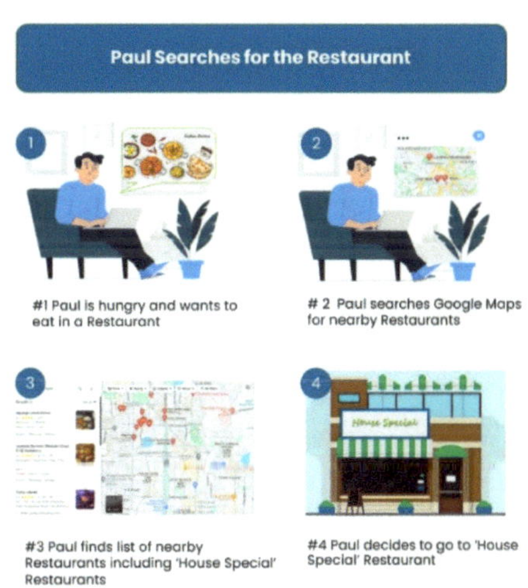

#1 Paul is hungry and wants to eat in a Restaurant

2 Paul searches Google Maps for nearby Restaurants

#3 Paul finds list of nearby Restaurants including 'House Special' Restaurants

#4 Paul decides to go to 'House Special' Restaurant

The HMI application searches for SCADA:

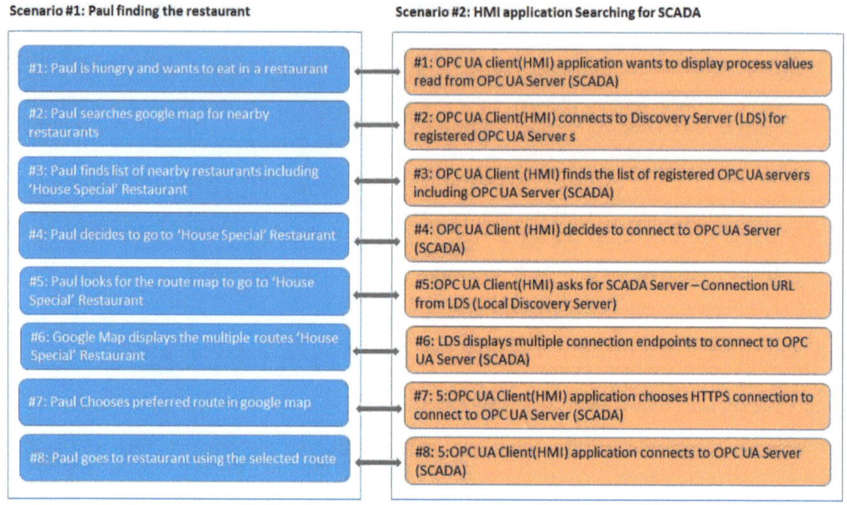

Mapping OPC UA Discovery Service to Paul's Search for Restaurant:

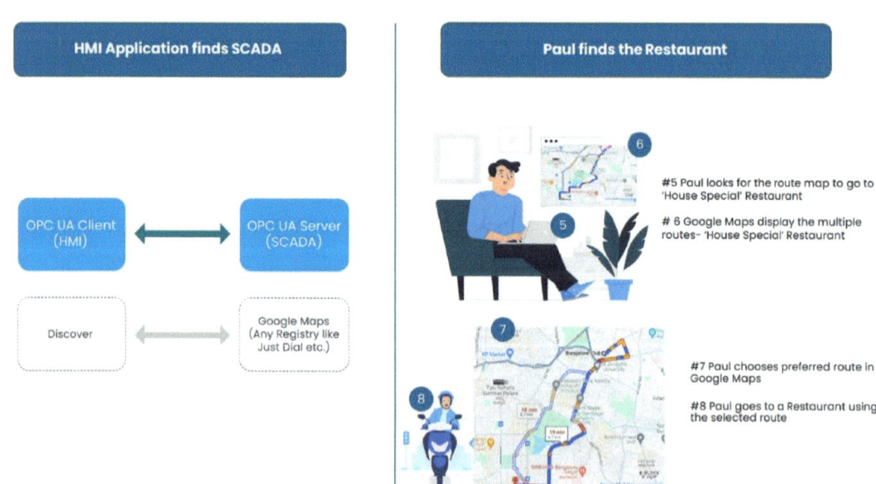

Route to the Restaurant mapped to OPC UA End Point

Opc.tcp://ABChost:4841/UtthungaOPCServer

Https://ABCHost:52521/UtthungaOPCServer

http://ABCHost:78689/UtthungaOPCServer

Understanding Discovery Service in OPC UA: Service

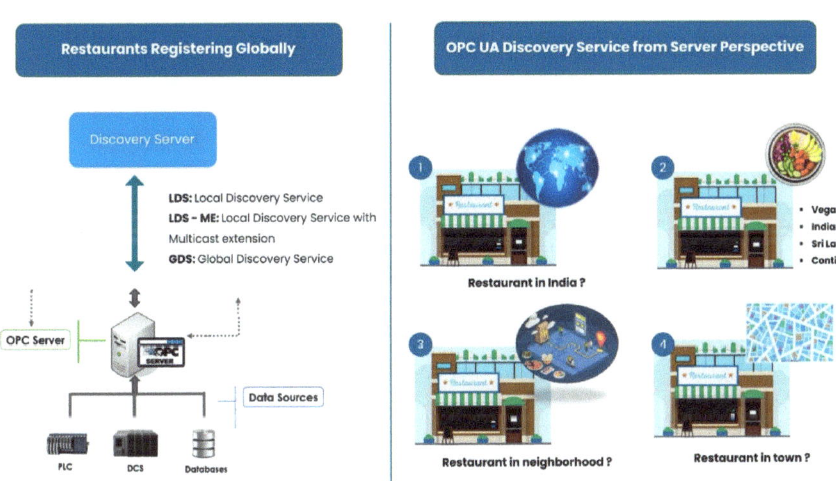

OPC UA Discovery Service from Client Perspective

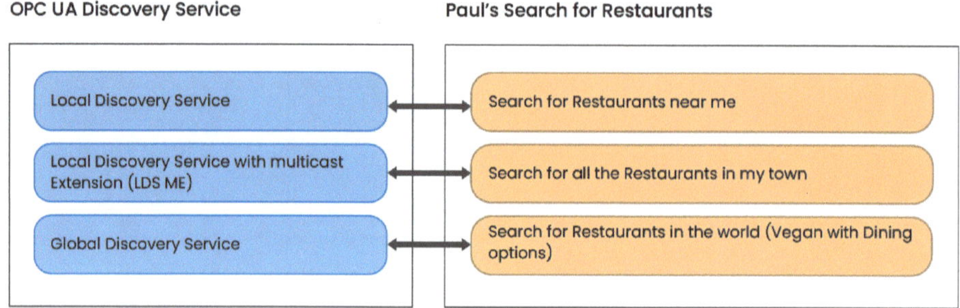

OPC UA Discovery Service from Server Perspective

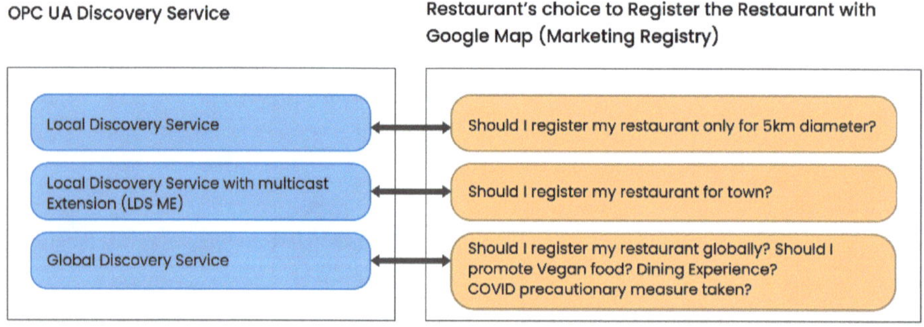

Once he's got a ride, he gets to the restaurant (the client establishes a session with the server) and wants to look at the menu (the address space) the restaurant is offering to figure out what he wants to eat.

'House Special' Menu Vs. OPC UA Address Space

When Paul looks at the menu (the address space) it's not just a long list of items. He can clearly see dishes segmented by whether they are a starter, salad, main course, drinks, etc. And the best thing! Subscription packages! He can choose any of the breads and salads for a monthly subscription and get free delivery (subscription). He can also personalize the menu for himself (create a view) and whenever he comes to the restaurant (Server) he gets only that menu and doesn't have to look at the other choices – saving him time and effort. Or he can state his preferences, like turn on the "veg only" option on a food delivery app and see only the menu items (variables) that fit those criteria.

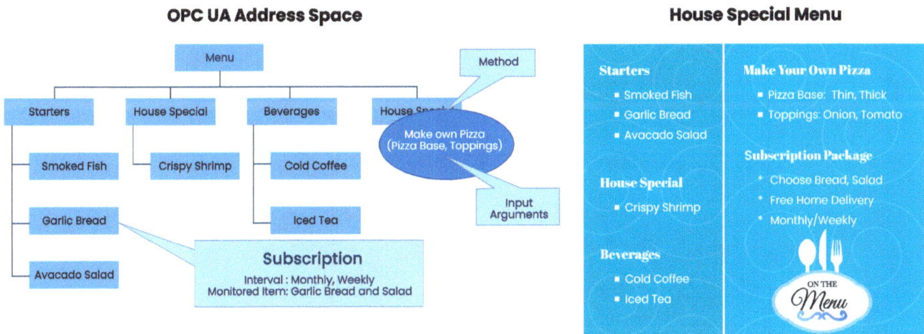

Illustration: OPC UA Address Space Similarlity to Restaurant Menu

Understanding Read/Write in OPC UA:

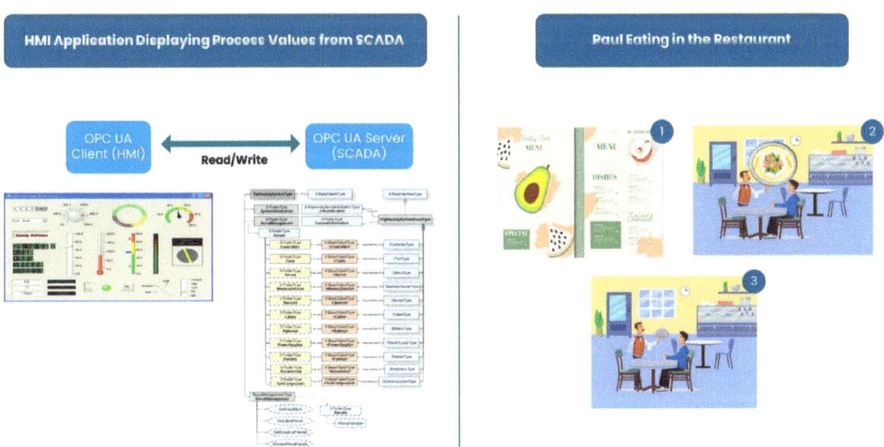

HMI Application Displays Process Values from SCADA:

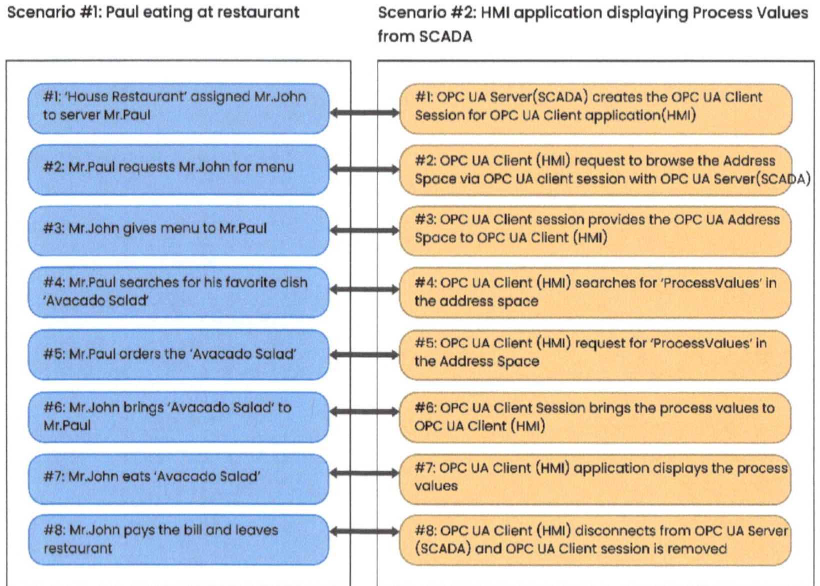

Understanding 'Subscription' in OPC UA:

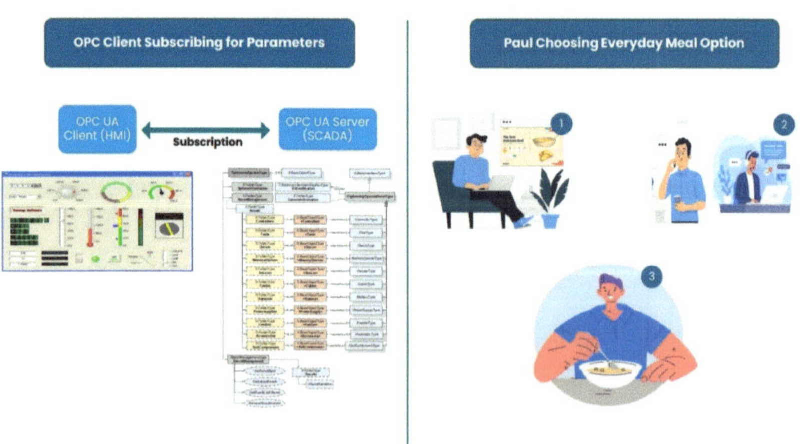

Paul Choosing the Everyday Meal - Mapping it to OPC Subscription Feature:

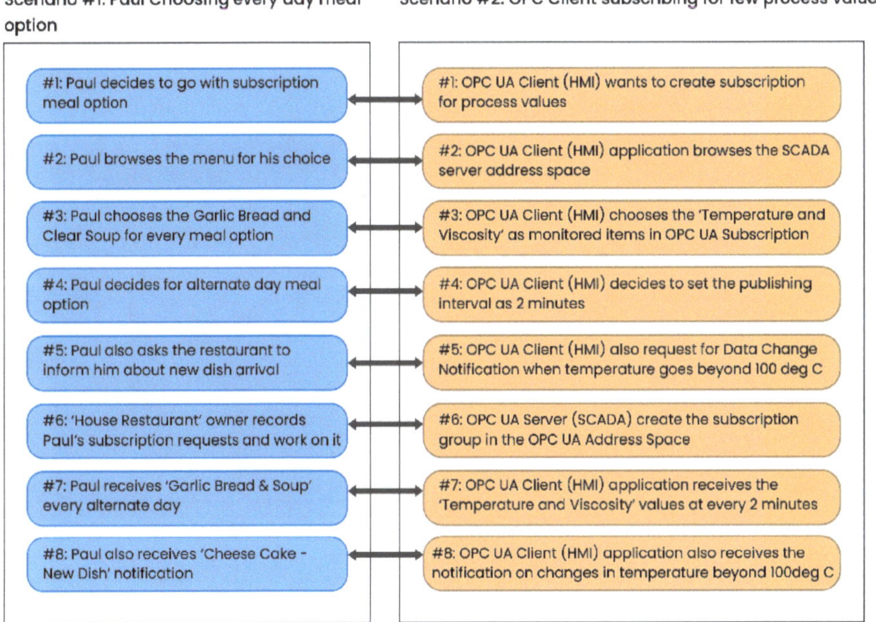

Understanding OPC UA in Layman Terms: OPC UA Pub/Sub Model

On another day, Paul ends up at a restaurant that doesn't serve a-la-carte food, but only a buffet. Which means, he can't order anything specific, but must eat from what is laid out on the buffet table. The table manager (broker) serves the food from what is available on the table, not just to Paul but to all the people (clients) who are in the restaurant.

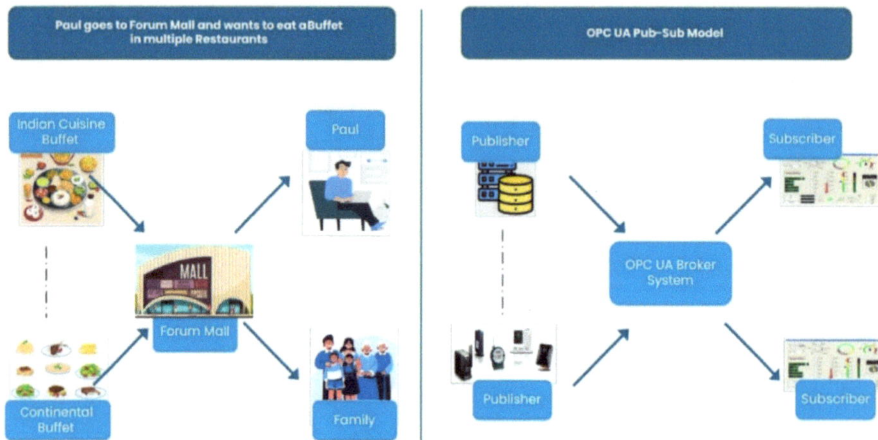

Once customers are satisfied, they can end their meal, easing the strain on the kitchen staff in a buffet setting compared to serving multiple customized orders. This approach simplifies managing numerous customers, as scaling becomes difficult with individual orders. In a buffet, the kitchen prepares a fixed selection and quantity of food, indifferent to the number of diners. Customers choose from what's available, and the table manager acts as a broker, serving client needs with the supplied options.

Paul Goes to Eat a Buffet

Now that we have seen how OPC UA client-server interaction can simplify things and enable IOT applications, let us explore how you can migrate to it from existing systems. But before that, a quick dive into OPC UA companion specifications.

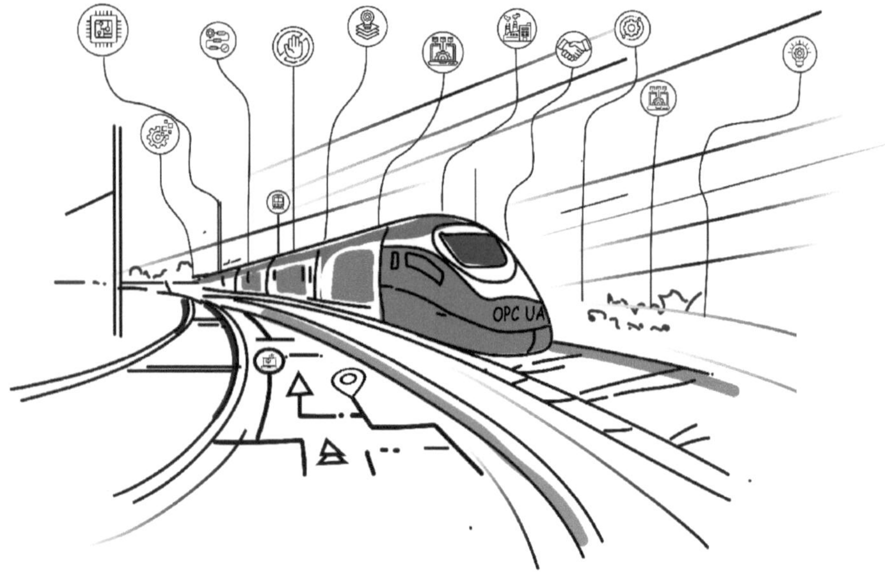
OPC UA

OPC UA Myth Busting #6

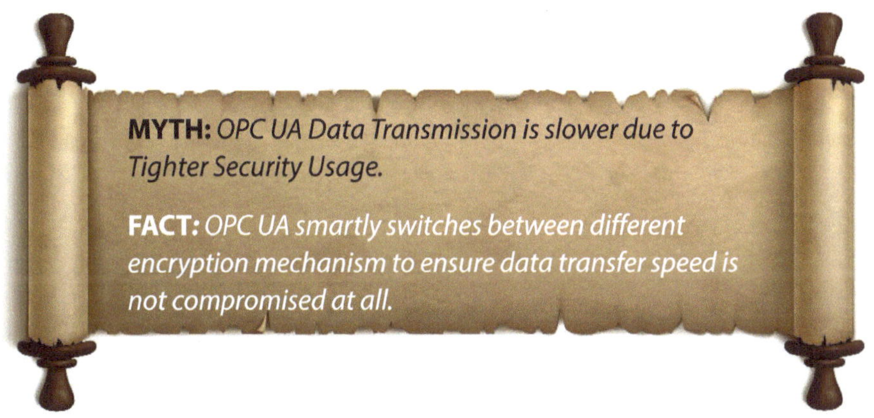

The myth that OPC UA's data transmission is slower due to its heightened security arises from a historically rooted belief about the trade-off between security and speed. While this belief might hold some truth in certain contexts, it doesn't apply universally, especially not to advanced systems like OPC UA.

Through its smart adoption of varying encryption mechanisms and the efficiency of modern encryption processes, OPC UA ensures that its "Industry Grade" security does not come at the cost of speed. OPC UA can dynamically switch between different encryption mechanisms based on the specific requirements of a situation. This adaptability ensures that while data remains secure, the speed of its transmission isn't compromised. Thus, users can benefit from the best of both worlds: robust security and fast data transmission.

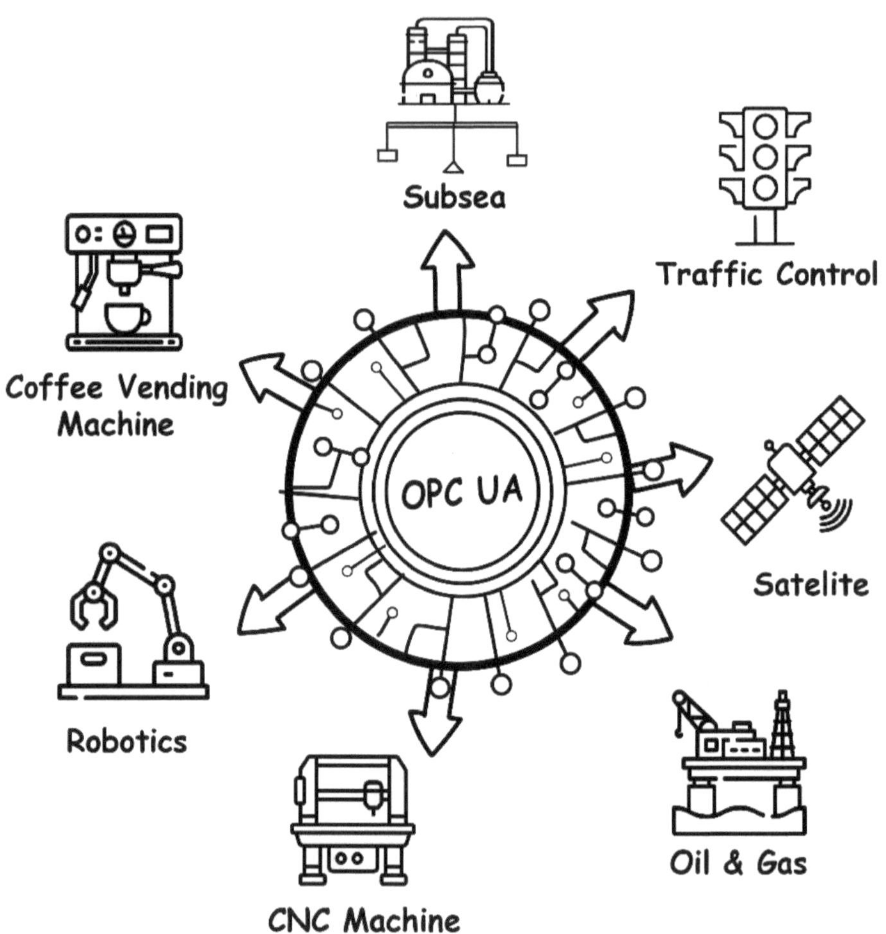

Subsea
Traffic Control
Coffee Vending Machine
OPC UA
Satelite
Robotics
Oil & Gas
CNC Machine

OPC UA & Companion Standards

In the OPC UA information model it's not just about transferring the data, but also the semantic information, or the meaning, of the data. To extend this capability to other domains, the OPC Foundation collaborates with other international associations (e.g., VDMA) and standard organizations and facilitates the development of OPC UA companion specification standards.

While OPC UA forms the foundation of communication, providing discovery, transport, information access, security, and robustness layers, Companion Specifications add another layer of specificity, tailoring semantics, behaviors, and data structures to meet industry or application requirements on top of this infrastructure. Leveraging the existing OPC UA Data Model and derived from OPC UA Base Information Models, like Data Access and Alarms & Conditions, Companion Specifications promote interoperability and standardization across systems. By enabling semantic-level interoperability, machines and systems can effectively communicate using standardized models and functionalities, bolstering seamless communication between disparate components.

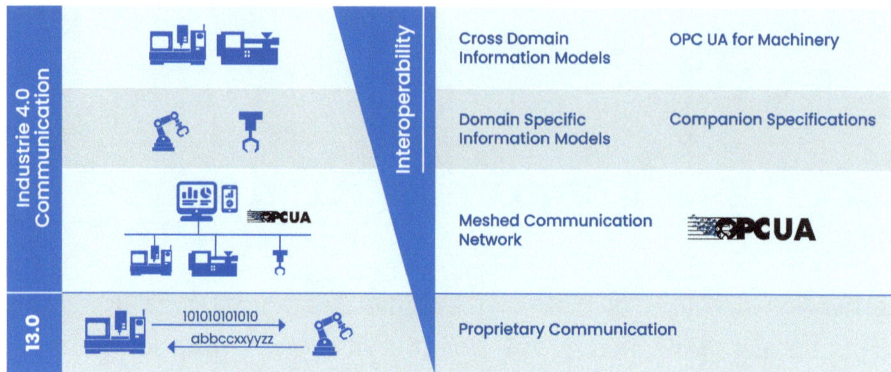

Reference: Illustration #15

The OPC UA companion standards bring about data contextualization. For instance, the OPC UA Companion Specification for Robotics provides a standardized information model which can present all robot-related data, regardless of manufacturer, type, or location, accurately. This companion standard covers all current and future robotic systems, such as Industrial robots, mobile robots, several control units, and even peripheral devices, which do not have their own OPC UA server. This companion standard also covers the asset management and condition monitoring use cases for these Robots.

Let's take an example of Robots exposing information like Motor Axis, Position, Identification etc., via OPC UA Server information model as shown below.

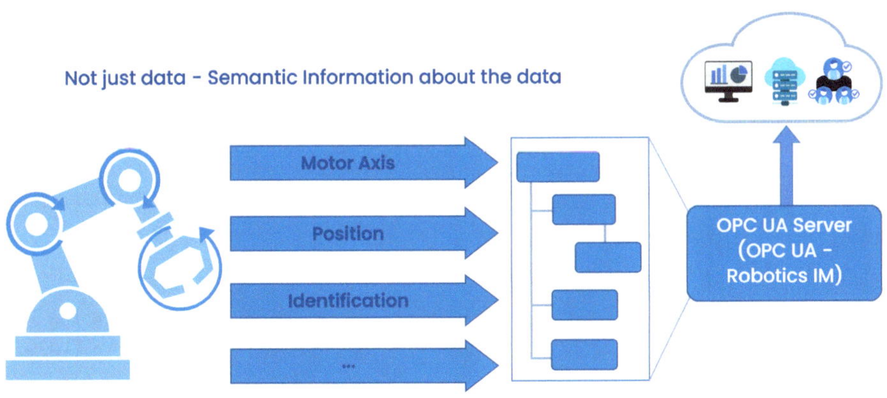

Reference: Illustration #16

The OPC UA client applications in the ERP system or in cloud can visualize the data provided by these robots for better analysis and decision making.

Companion standards also help bring more uniqueness via vendor specific information models.

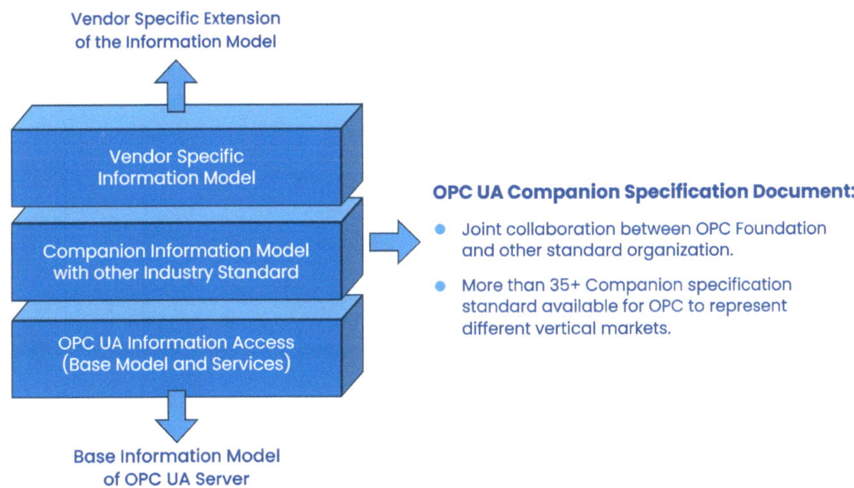

Reference: Illustration #17

Consider a scenario in an automation plant, where an ERP system running OPC UA Client fetches the information from assets in the field for plant operations. This plant has robots from Vendor A and Vendor B. Vendor A's Robot is a basic robot, whereas Vendor B's Robot is more advanced version and has support for sophisticated features like 'Fault Recovery Mode', which is not available in Vendor A's Robot.

From the OPC UA perspective, both Vendor A and Vendor B Robots expose information like 'Motor Axis and Position' via the Robotics companion information model. However, on top of this, Vendor B's robot exposes the 'Fault Recovery Mode' via their specific information model.

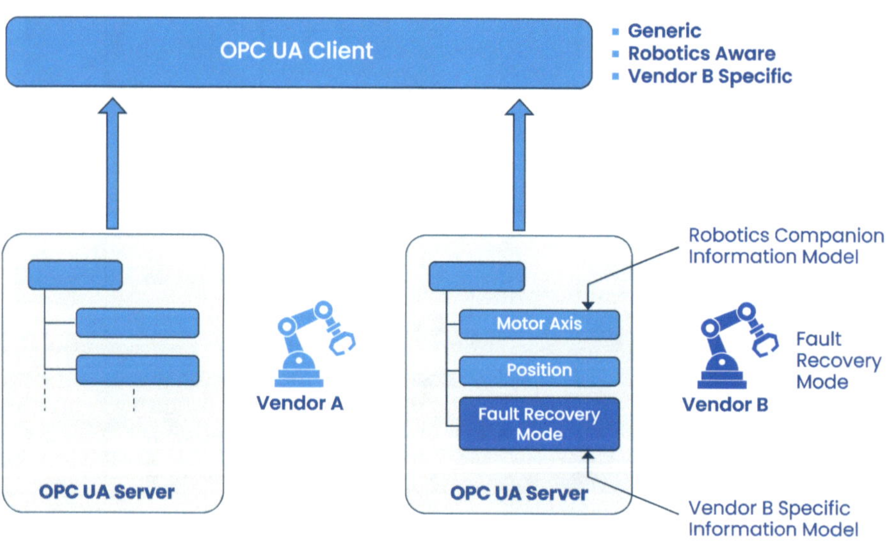

Reference: Illustration #18

Some of the OPC UA Companion standards include:

Vertical Market/ Domain	Companion Standards
Automation	PLCOpen, AutoID Supply Chain, Plastics and rubber machinery, Machine Vision, Robotics, Powertrain, Weighing, Woodworking, Casting, End-of-Arm-Tools, Vacuum Pumps, Glass Industries, Mining, Compressed Air Systems, Tobacco, CNC, Food Beverage, Commercial Kitchen Industry, Functional Safety, Engineering, Open-SCS Pharmaceutical etc. AssetManagement
Automation - Process Industry	FDI, FDT OPC UA, PA-DIM, DI, ADI etc.
Automation - Fieldbus	EtherCAT, PowerLink, PROFINET, CIP, Sercos, IO-Link, CAN in Automation, CC-Link, BACNet for Building Automation, IEC61850 - Electrical substation etc.
Oil & Gas	MDIS, Standard Leadership Council (SLC), DSATS (Drilling System Automation Technical Section), Energistics
Standardization	ISA, W3C, IDS (Industrial Data Space), OpenFog etc.

Note: Refer OPC Foundation website for more updates on companion standards.

Standardized information models, such as OPC UA Companion Specifications, enable companies to fully leverage Industry 4.0 and the IIoT.

As part of the Study on Interoperability in Mechanical and Plant Engineering published in 2021, the VDMA surveyed more than 700 companies about OPC UA and interoperability. The survey results indicate that companies assign a high level of relevance to OPC UA for different reasons. Some 90 percent of the surveyed companies have already implemented OPC UA as an interface or plan to do so in the future. However, this does not mean that the goal of interoperability has already been achieved – this is only possible if the participating communication partners also understand the interface contents.

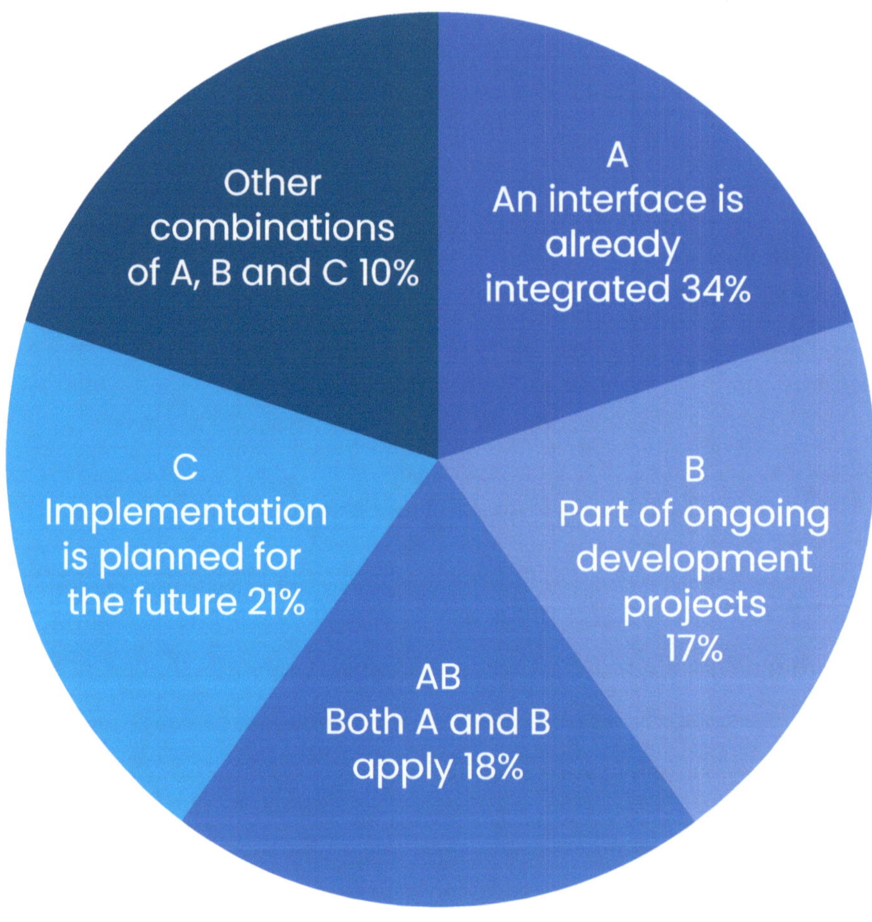

Reference: Illustration #19

Developing OPC UA Companion Specifications

To meet the global market pressures companies and users must reduce non-value-adding processes. There is also a need to reinvent interfaces, reduce the high efforts in integration projects and in using the information in post-processing with applications, and the mandate to comply with some standard. To reduce these efforts, it makes sense for companies to collaborate and create companion specifications for OPC UA to bring in standardization and reduce efforts.

Standardized OPC UA companion specifications can be used as base parts of interfaces, facilitate integration processes, and make information usage much more accurate and accountable. Creating companion specifications gets companies closer to compliance and reduce non-value-add efforts.

Illustration: Companion Specification Creation Workflow

Step 1: Establish a Working Group

Developing such a standardized information model requires a careful, iterative process along with the expertise of different stakeholders. Therefore, a working group to develop companion specifications must include industry experts, machine manufacturers, end users, and editors who are intimately familiar with OPC UA and can take on the modeling tasks.

Step 2: Develop Specific Use- cases to Determine the Functionalities Needed

Once such a committee is formed, the actual need for functionalities is determined at the outset. This is done by identifying and describing specific use cases to be addressed by the Companion Specification. This seemingly subordinate step is important. The result of this step will provide a clear and precise objective and guidelines for the entire development process of the information model.

Step 3: Compile All Elements Necessary to Fulfil the Use-case Requirements

Once the use cases have been developed, the working group begins to compile all the elements necessary for their fulfillment. For example, this may involve the mapping of individual measured values or parameters. Apart from this method calls, representation of machine & component statuses, and a holistic mapping of production orders are also of great significance.

Step 4: Develop the Information Model

Based on this preliminary work, the most extensive and time-consuming part of the development process of a companion specification begins - I.e., the development of the information model.

This includes the development of semantics to enable uniform and comprehensible communication between the systems. This is a complex work step, which often requires intensive discussions and coordination in the working group. The data structures that are relevant for communication and data exchange are also determined while developing the information model. It is critical that the models are well thought out and flexibly designed to meet the different requirements of the various use cases.

Since errors in modeling semantics, designing references, or choosing data types of individual objects can cause serious problems in practical implementation, it is advisable to perform these topics with utmost care and expert support. There are many solution approaches that are correct in the technical conversion but are undesired or even harmful in practice.

A simple example of this is the incorrect application of the data type "Enumeration". In many cases, it is meaningful to specify a selection of values firmly, which can be converted by this data type. However, there is no possibility to extend this selection of values afterward, they are unchangeable values. In case that after the publication of the Companion Specification the requirements change in such a way that an extension is necessary, the published Companion Specification can no longer be implemented. To be able to fix this error afterward, a re-establishment or new formation of a suitable working group is required. Indeed, changing a data type requires a so-called "breaking change", which is equivalent to a complete republication of a revised standard. The potential danger of such modeling errors can only be minimized by exchanging experiences, observing best practices, and skillful modeling.

Step 5: Draft the Companion Specification

Next, the working group drafts the Companion Specification. This draft goes through several review and validation loops to identify potential weaknesses and validate the information model. The input of the various applied-industry experts is invaluable here to ensure a high level of quality and realism. After successful review the draft is submitted to the public for comment via the OPC Foundation and VDMA (for OPC UA Companion Specifications developed by the VDMA). External feedback and suggestions are incorporated into the development process and help to address any aspects that may have been overlooked. Once the final version of the Companion Specification is completed, it is published and available to the specific industry, and the OPC UA community. However, this step does not mark the end of the process, as the working group remains responsible for maintaining, updating, and extending the Companion Specification even after publication. In this way, developments in the industry can be taken up, and the benefits for users can be continuously improved.

A good example of the implementation of the described approach can be found in the OPC UA for Tightening Systems Companion Specification (VDMA 40451-1). This document specifies an OPC UA information model that represents the controller, tools, servos, sensors, and some other subcomponents of a tightening system.

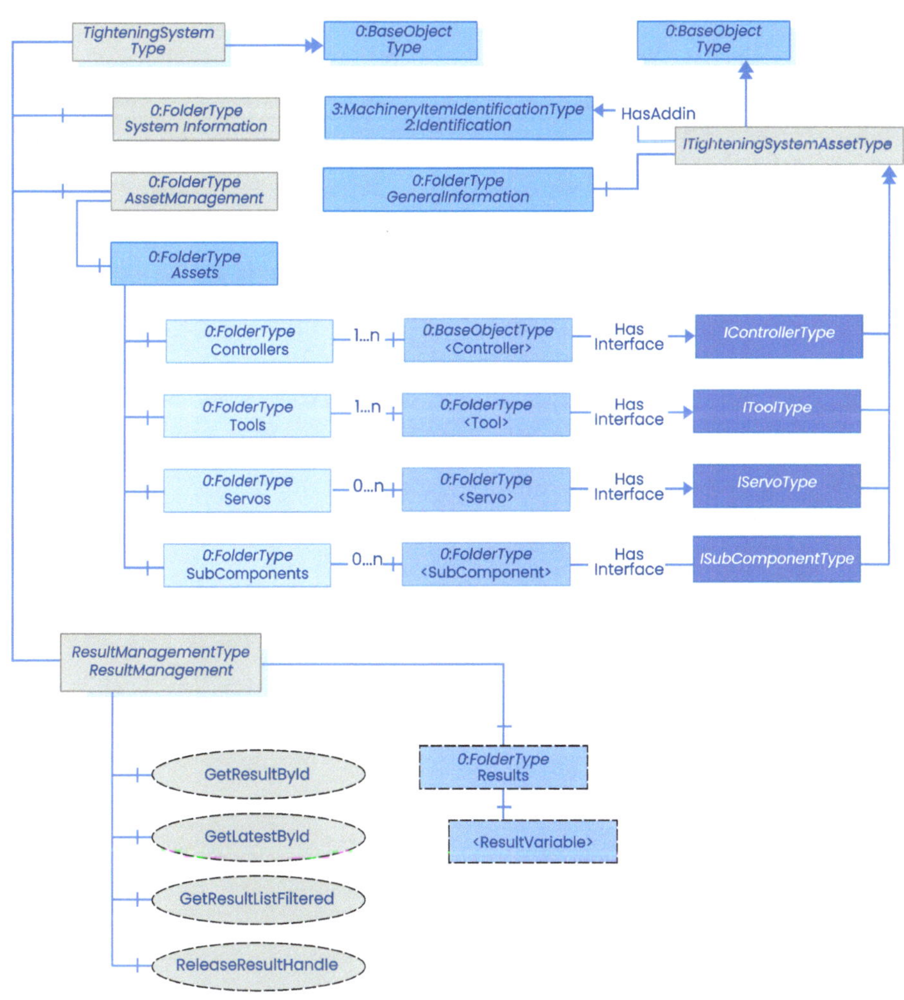

Illustration: Information Model for Tightening Systems

The figure shows an overview of the information model of such a tightening system. The so-called TighteningSystemType has the components SystemInformation, AssetManagement, as well as ResultTransfer. In this standard, the AssetManagement contains the mapping of the physical components of such a machine. More detailed data on the individual components are located within the objects marked in yellow and are not shown on this overview. The references shown between the yellow folder structure and the associated object marked in orange are "mandatory -" and "optional placeholders". With the help of these structures, the end user of the Companion Specification can create the individual components partially, not at all, once, or in arbitrary number. This clarifies that with professional interpretation of such a standardized information model the application is applicable to complex systems, as well as to smaller solutions. In this case, this means that a tightening system with only one controller and one tool can be implemented in the same way as a tightening system with, for example, three controllers, seven tools and seven servo drives.

Machine-related data, such as machine identification, is mapped via the SystemInformation, while ResultManagement - implemented by several method calls - enables structured access to created data sets. However, such use cases as machine and component identification or the mechanism of individual data transfer of data sets are not only interesting in the Companion Specification for Tightening Systems. Some such industry-independent requirements represent an essential component for a wide variety of machines across all industries.

Harmonization of Companion Specifications is Necessary to Enable Interoperability

To avoid disparate approaches and ensure consistency across various companion specifications, a significant number of parameters, functionalities, and features have been identified as universally applicable across different industries and domains. These include essential aspects such as Identification (Serial Number, Manufacturers, etc.) and Job Management.

To address this harmonization requirement effectively, the concept of creating a base specification has emerged. This base specification aims to serve as a comprehensive framework covering these universal topics, providing a common ground for all industries to build upon. By establishing a harmonized foundation, industries can streamline their implementations, foster interoperability, and promote a seamless exchange of data and information, ultimately enhancing efficiency and collaboration across the entire ecosystem.

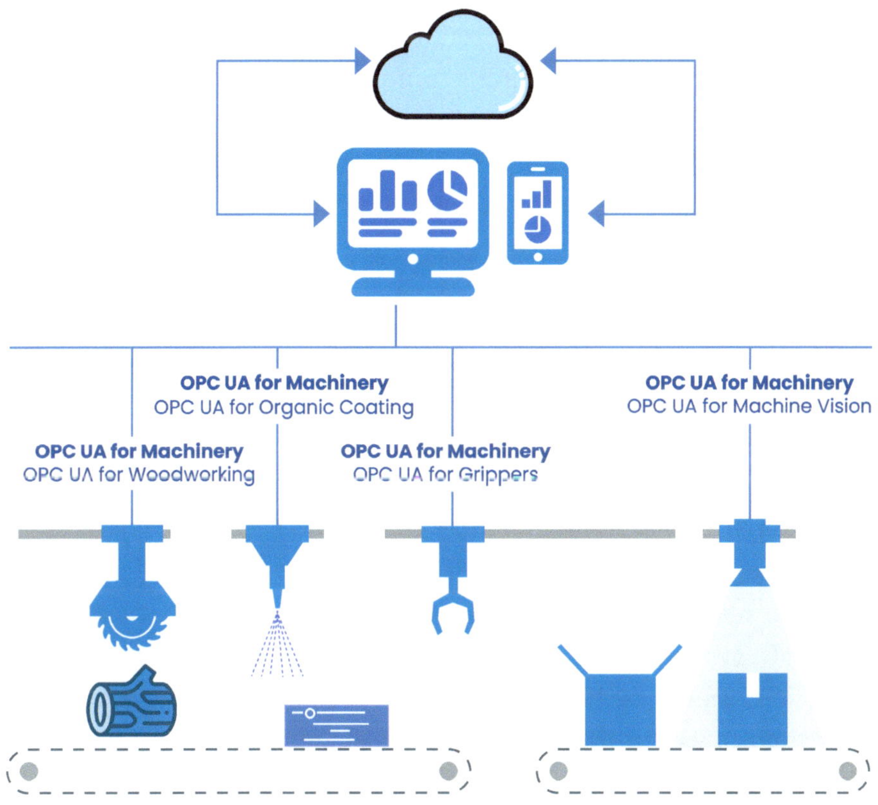

Reference: Illustration #20

OPC UA for Machinery:

The OPC UA Companion Specification "OPC UA for Machinery" is a key element in achieving the highest level of interoperability through cross-domain information models. It comprises various building blocks for machinery, facilitating diverse use cases across different types of machines and components. These building blocks can be utilized either in other companion specifications or directly within an information model. Initially published in September 2020, it was one of the pioneering OPC UA companion specifications fully endorsed by the **UMATI** initiative. Recently updated to accommodate additional use cases, it continues to undergo further enhancements. The specification encompasses crucial building blocks such as Machine Identification, Component Identification, Machinery State, Counters, Process Values, Job Management, Result Transfer, and Energy Management.

As an example, the Identification section of "OPC UA for Machinery" focuses on the crucial task of giving each machine a unique identity. Creating this identity is essential when dealing with various OPC UA Servers or combining them. This section helps create a standard way to know important things about each machine, like who made it and its serial number. Such unique identifications help easy tracking within the OPC UA system. Users can get helpful information about the machines, empowering them to use the equipment more effectively. It also allows OPC UA Clients to add specific information for each machine, making them more flexible to customize. This robust identification system is vital for smooth communication and working together between different machines in industrial settings. The *MachineryItemIdentificationType* is a primary type that is used by both *MachineIdentificationType* and *ComponentIdentificationType*.

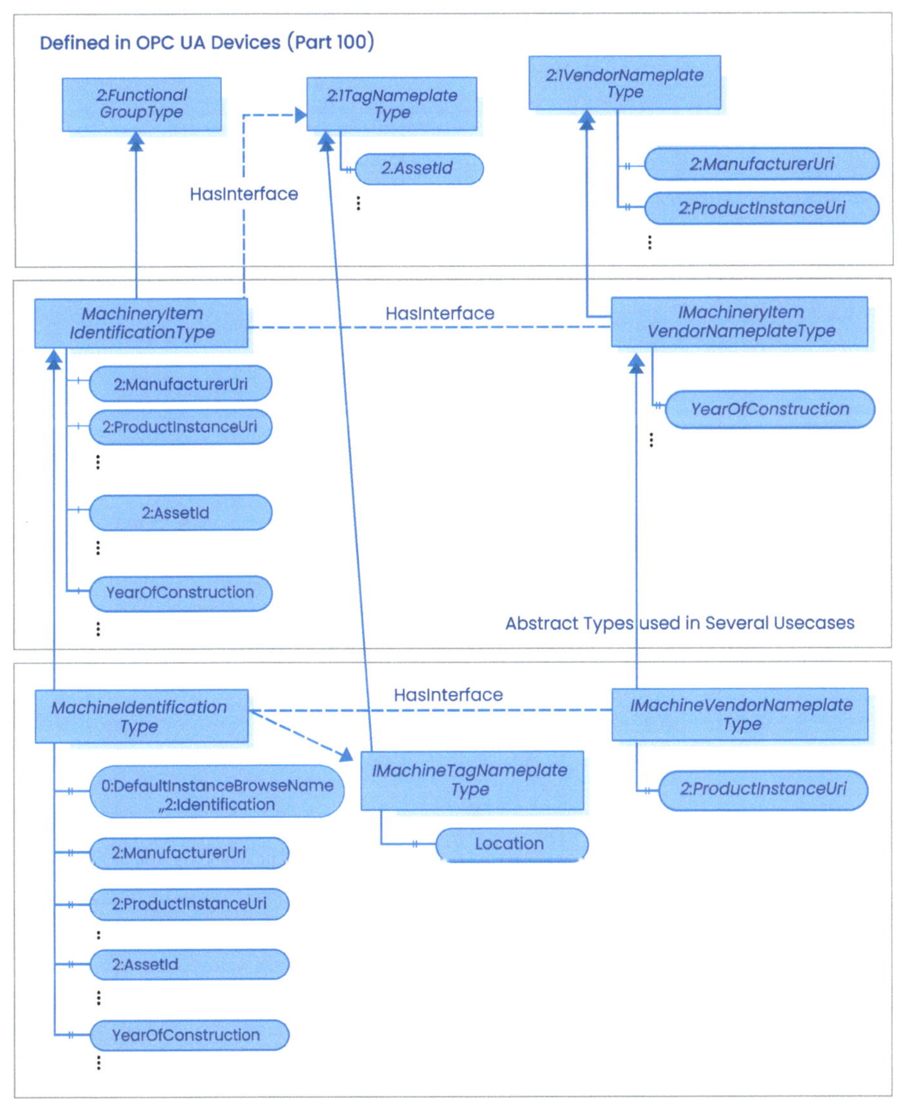

Reference # 21: Machine Identification Building Block

Another Building Block, which has aroused great interest across industries for its release in mid-2023, is the **OPC UA for Machinery - Part 3: Job Management**. This Companion Specification developed in collaboration with the ISA-95 Job Control Working Group enables crucial tasks such as transmitting work orders, checking their status, controlling them if necessary, and deleting the data from the machine once the job is completed. The following figure shows the Machinery Building Block Job Management information model.

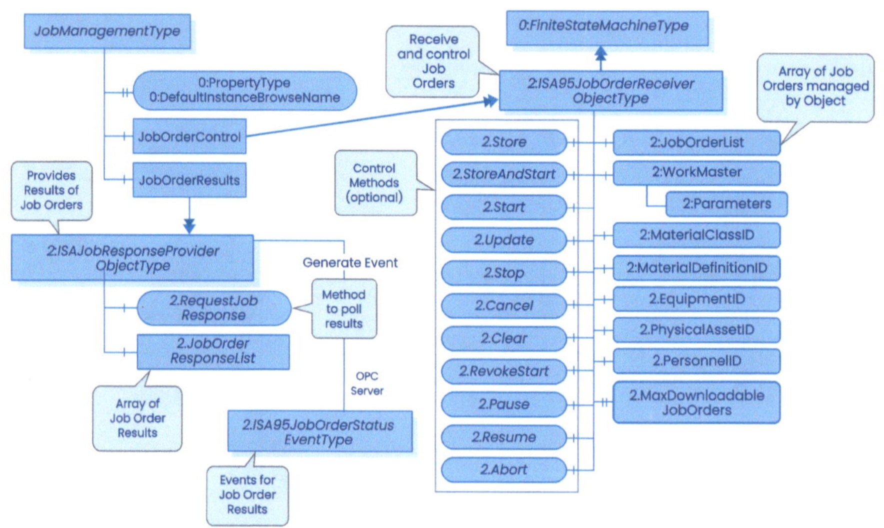

Illustration: Job Management Building Block

The parent Job ManagementType has two components JobOrderControl and JobOrderResult. Here, the 'type' definition is described by the ISA95JobOrderReceiverObjectType, as well as the ISA95JobOrderResponseProviderObjectType.

To facilitate fully comprehensive order control, numerous methods are available under the ISA95JobOrderReceiverObjectType. An array of orders can be represented via the JobOrderList, whereby the order of the respective orders is particularly important. The array starts with all the executed orders, followed by the currently executed or cancelled orders, orders awaiting to be executed, and finally, the orders that cannot be executed.

The ISA95JobOrderResponseProviderObjectType represents the report about executing an order. For example, it provides the results of the job, an array of results reports, or the ability to request specific results from the machine.

To better understand the details, like the parameters you can send to a machine for a job, it's a good idea to study the **OPC UA for Machinery - Part 3: Job Management (VDMA 40001-3)** standard. It is important to note that the other sections in this publication have also provided significant value-add in making the job easier for many working groups. One can get all these standards for free from the VDMA and OPC Foundation websites.

One of the standouts features of OPC UA for Machinery lies in its implemented use cases for machine and component identification, forming the foundation for the "Plug & Work" concept. This enables the standardized recognition of diverse network participants, promoting machine- and manufacturer-independent communication in various industries. The "Machinery State" module enables valuable use cases such as Machine Monitoring and KPI Calculations, contributing to increased machine availability and efficiency, especially in production setups that involve machines from different industries.

While OPC UA is the preferred technology for Industry 4.0 communication in production, OPC UA for Machinery serves as the fundamental specification for machinery and plant engineering. It provides information models with validity across numerous machines and plants, fostering cross-industry interoperability. The base specification has paved the way for modern, networked manufacturing, making it an ideal entry point into the world of OPC UA technology.

A Strong Network of Companies and Organization is Driving the Global Implementation

Currently, approximately 40 working groups within the VDMA are developing such companion specifications, involving over 700 companies and industry users. The acceptance and adoption of OPC UA relies on a critical mass of participating companies. The industrial SME sector benefits from such a standard as it reduces the costs of developing customer-specific interfaces. The barriers to integrating machines into existing and new systems are lowered. Machines and systems can interact with each other flexibly, efficiently, and resource-efficiently (referred to as sustainability). By establishing OPC UA as the guiding standard, the VDMA ensures that machines and systems can communicate in a unified language. This forms the basis for the "Plug & Work" promise.

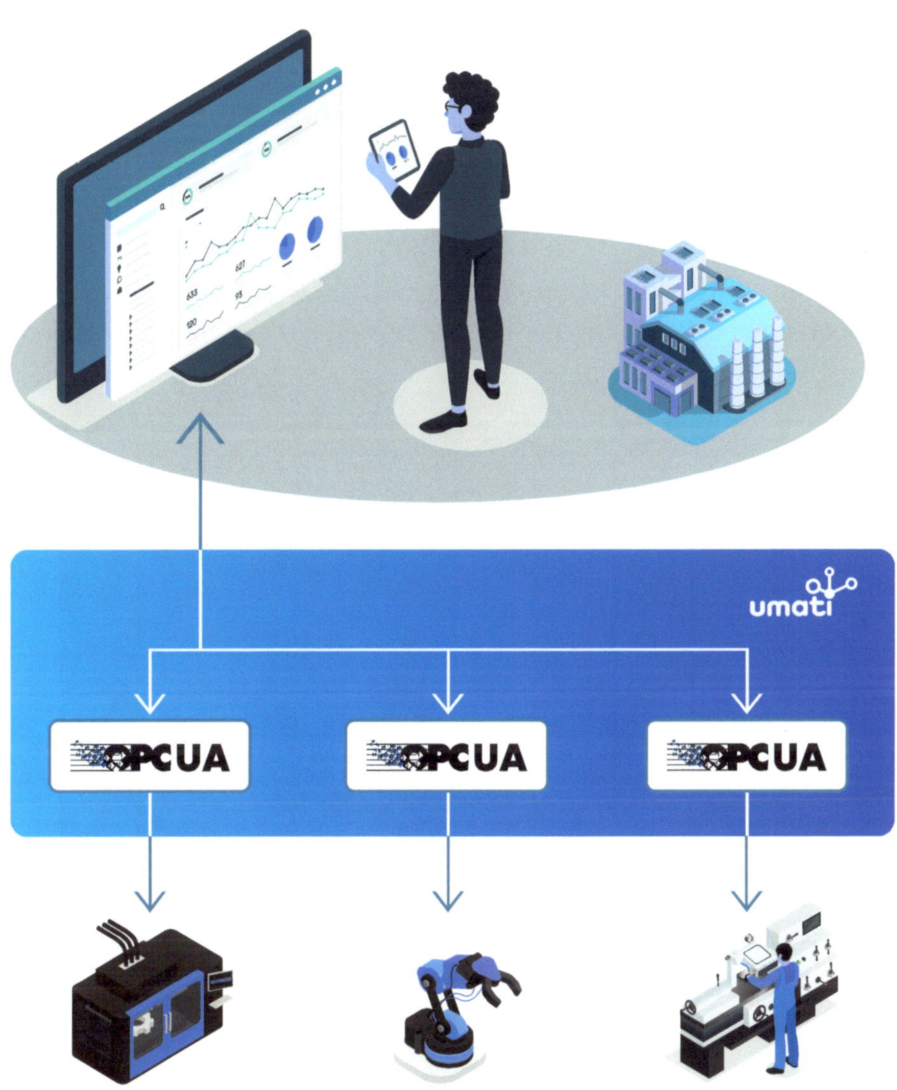

Illustration: UMATI Supports Implementation of
Companion Specifications *[UMATI.org]*

The UMATI (Universal Machine Tool Interface) initiative promotes
the worldwide adoption of companion specifications to foster global

interoperability. More than 300 partner companies and institutions have joined forces in the UMATI implementation initiative to collectively advance the practical application of OPC UA companion specifications. In addition to community and implementation support, an open dashboard serves as a demonstrator. It provides a reference architecture with a recommended OPC UA configuration. Furthermore, UMATI supports the global dissemination of OPC UA companion specifications as the leading standard for interoperability through strong branding.

OPC UA
OPC UA
OPC UA
OPC UA
SUCCESS
DATA
ANALYSIS
OPTIMIZATION
OPC UA

OPC UA Myth Busting #7

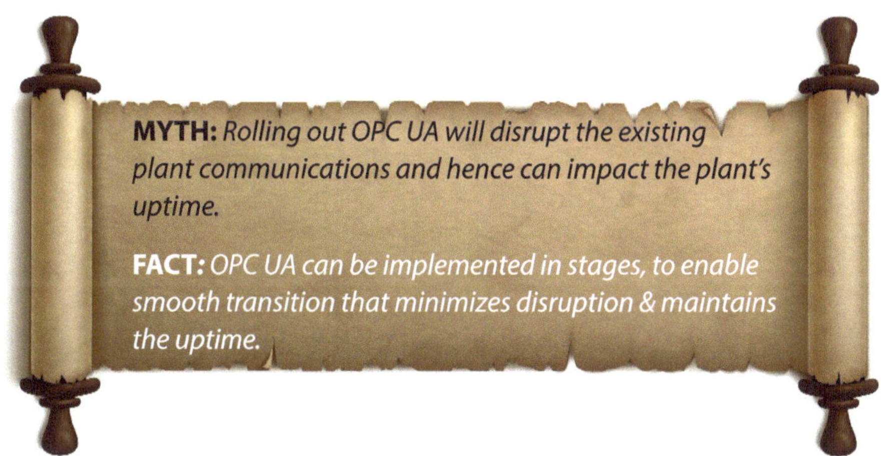

The myth that implementing OPC UA disrupts existing plant communications and impacts uptime likely stems from general concerns about introducing new technology in the existing environment. Such concerns are valid since any major change to a plant's communication protocol in an industrial environment can seem risky. This belief may have its roots in past issues with older, less adaptable protocols, which were indeed disruptive when updated or replaced. Although OPC UA was a significant innovation over OPC Classic, there is a misconception that it cannot integrate with OPC Classic or other legacy communication systems.

This misconception about OPC UA arises from a lack of information. Industry experts and the OPC Foundation have periodically provided guidelines on migrating from legacy setups to OPC UA-based communication. Developed to address limitations in previous automation protocols, OPC UA improves interoperability, security, and scalability. Its compatibility allows integration with existing systems, regardless of their communication protocols. Thus, OPC UA enhances rather than replaces existing infrastructures, adding functionalities, and improving communication without disrupting current systems.

Therefore, in most cases, the implementing OPC UA can be achieved in stages. This allows smooth transition, minimizes disruption, and maintains maximum uptime.

• • •

OPC UA
OPC DA

Migrating to OPC UA

In a competitive landscape, the window to move to OPC UA is shrinking, creating a sense of urgency for industrial players. However, most organizations and business leaders hesitate to act because they don't know how to go about this shift. In this section, we will explore how you can embrace OPC UA and all its possibilities with minimal disruption.

MIGRATING FROM OPC DA to OPC UA

The popularity of OPC Classic and its widespread adoption meant that by the time the OPC Foundation came up with OPC UA, most companies were already running on OPC DA. Shifting to OPC UA, cspccially for larger plants, could take several years. And therefore, there was a need for a seamless mechanism to connect OPC UA to OPC classic and vice versa.

However, the shift was inevitable. The drawbacks of OPC DA were creating challenges for IT teams, system integrators (SI), and end customers. For instance, COM/DCOM uses dynamic ports for making calls across servers and these ports cannot be defined. Hence, to use OPC DA IT teams had to open multiple ports in the firewall. OPC servers also required numerous COM permissions to be open within the machine. This made the system prone to cyber-attacks. From an SI and end customer perspective, using OPC DA was problematic in other ways. For example, OPC Classic Servers are incompatible with current OS software versions. This means that every new major upgrade to OS breaks the existing running OPC classic servers and requires reconfiguration. In addition, end suppliers provide limited support for OPC Classic servers making it hard to acquire defect fixes and feature releases.

While companies understand this need to make the transition to OPC UA, they do have some concerns:

- How do we embark on a phased and well-planned upgrade from OPC Classic to OPC UA communication?

- How to reduce the risk of unknowns and communication failures?

- How to ensure minimum downtime or data disruption while executing such a phased rollout?

- How to obtain proof of concept before rolling out a new phase of OPC UA?

Three key tools help us address these concerns - OPC UA Proxy, OPC UA Wrapper, and OPC UA Tunneller. Let's look at each of them.

An OPC UA Proxy is software with an inbuilt OPC classic Server and UA client. So, it enables an OPC classic client to talk to an OPC UA server (See Fig).

Illustration: OPC UA Proxy Architecture

An OPC UA Wrapper acts the opposite of a Proxy. It has an inbuilt OPC UA server and a Classic client. So, it enables an OPC UA client to fetch data from an OPC classic server (See Fig).

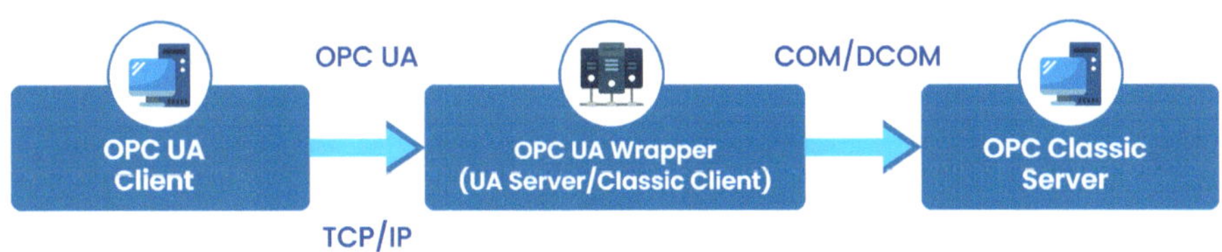

Illustration: OPC UA Wrapper Architecture

An **OPC UA Tunneller** is a combination of UA Proxy and UA Wrapper working together across the network firewall. This effectively converts the network traffic to OPC UA TCP/IP from the native COM-DCOM, as shown below. These TCP/IP settings are easy to configure in the firewall using a single port.

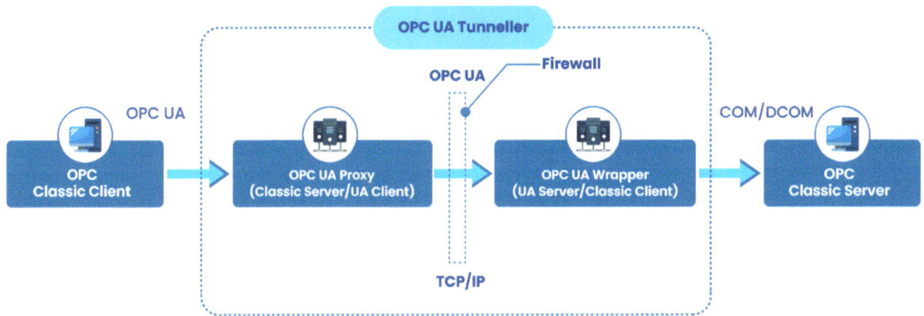

Illustration: OPC UA Tunneller Solution

Let's explore how this migration happens with a live example that we executed for a client. So, our customer was running an OPC Classic client setup with the following components:

- 3 OPC Clients running in the Office Domain (OD)

- 3 OPC Server running in the Process Control Domain (PCD)

- Every client is connected to every other server

- These clients/servers are connecting across DMZ, which is located between Office Domain (OD) and Process Control Domain (PCD)

- Multiple COM ports are open in the firewall to enable these communications

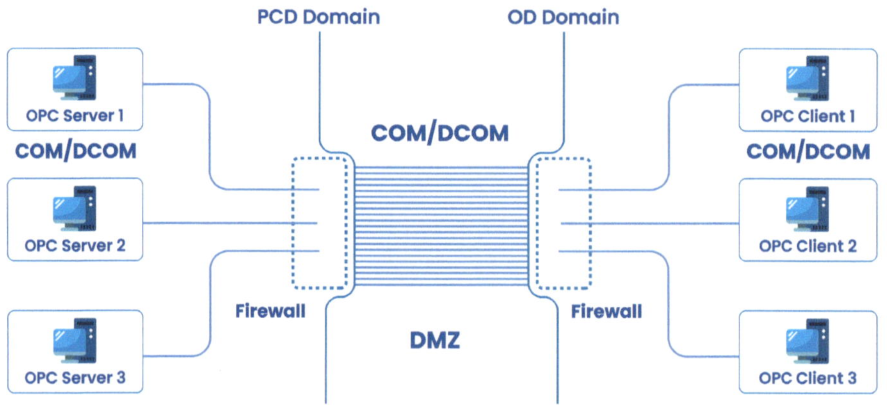

Illustration: Existing OPC Classic Setup Using DCOM Across DMZ

While the diagram shows only a few lines (representing ports) in the DMZ zone, the fact is that these lines number in thousands as the COM-DCOM demands a range of ports.

Step 1: Clean up the OPC Classic Communication (COM/DCOM) at the Network-level with the Help of OPC UA Tunneller

The first step is to introduce an OPC UA Tunneller between the OPC Classic Clients and OPC Classic Servers. This will ensure the following:

- The DMZ environment will have to open only one firewall port

- The COM/DCOM communication, which is network hungry, will be replaced by the highly optimized OPC UA communication at the network level

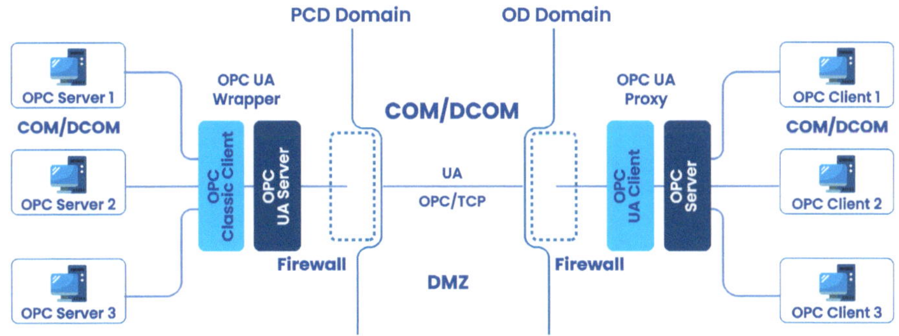

Illustration: DCOM Bypass for OPC Classic Setup Using OPC UA Tunneller

Once the OPC Classic communication has been removed from the network layer and DMZ environment, it provides conceptual proof that OPC UA communication is working at the network level, and the team also learns the required configuration. As the OPC Classic and COM-DCOM are removed, the IT team can close the ports, keeping only a few open for OPC UA TCP/IP communication.

Step 2: Start replacing OPC classic components with OPC UA components

Now that the troublesome DCOM communication is handled, the next step would be to replace the OPC classic components with OPC UA ones. For this customer, we did this replacement in two different phases. In both phases, the setup was first run with the existing classic components in redundant mode. Once the newly configured OPC UA components were found adequate, the classic components were taken offline.

Phase 1 - Provided OPC UA clients to replace the old classic clients

- 3 OPC UA clients were installed in the OD domain in parallel with classic OPC clients

- These 3 OPC UA Clients can connect to the Wrapper components directly to enable the OPC UA communication

- Now, we have 3 OPC Classic clients and 3 OPC UA clients running concurrently

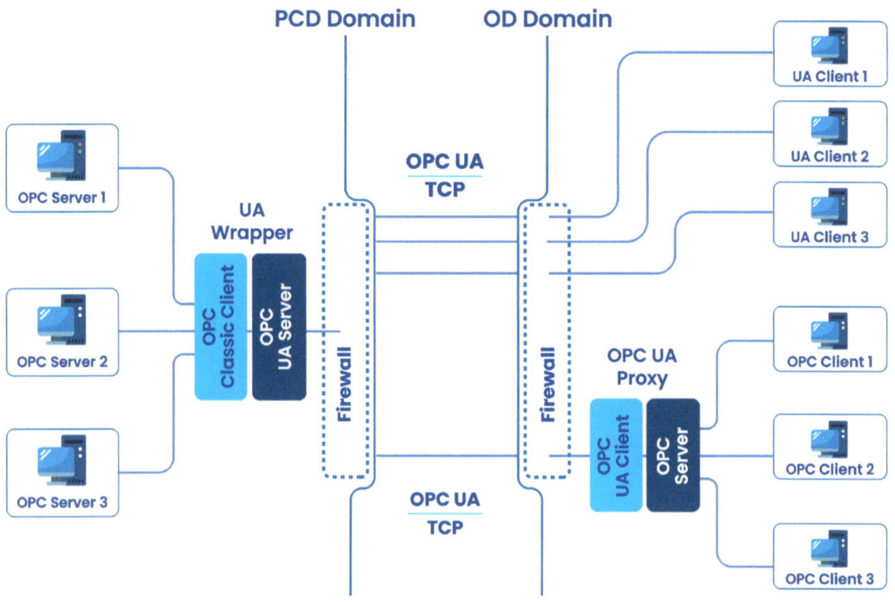

Illustration: Replacing Some of Classic OPC Clients with OPC UA clients

The setup configuration after removing the OPC classic client looks as below.

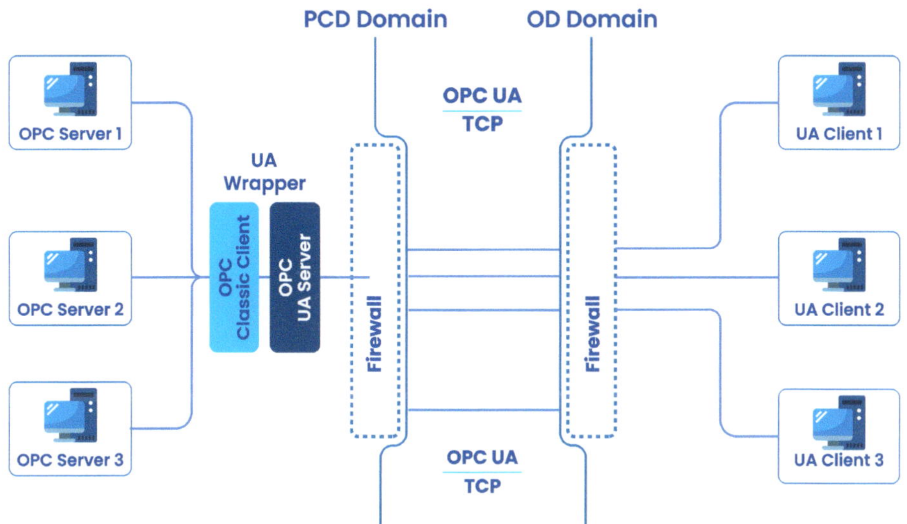

Illustration: Replacing All Classic Clients with OPC UA Clients

Phase 2 – Provided OPC UA Servers to Replace Existing OPC Classic Servers

- Install the OPC UA servers in the same PCD network as the OPC Classic server and get data from the same source

- Configure the OPC UA clients to pull data from the OPC Classic server via wrapper and from the newly installed OPC UA servers

- Compare the data coming from both servers and make sure they match all the expected parameters and give the same or better performance than the old classic Server

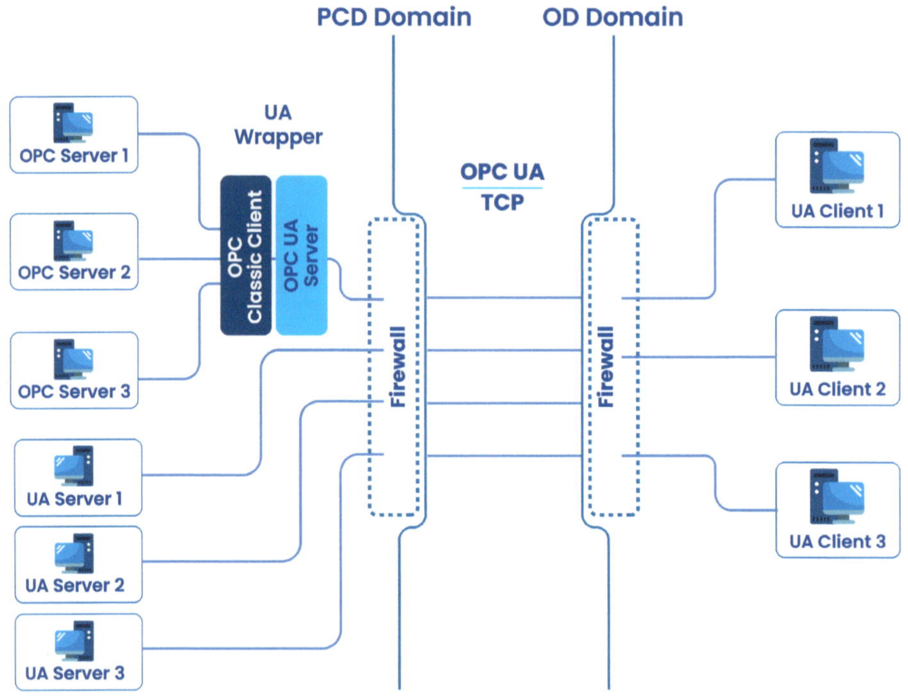

Illustration: Replacing OPC Classic Servers with OPC UA Servers

Once the OPC UA server communication is stable and verified, the UA Wrapper and OPC Classic server can be deprecated. Now the complete OPC Classic setup has been upgraded to OPC UA server, as shown in the below architecture.

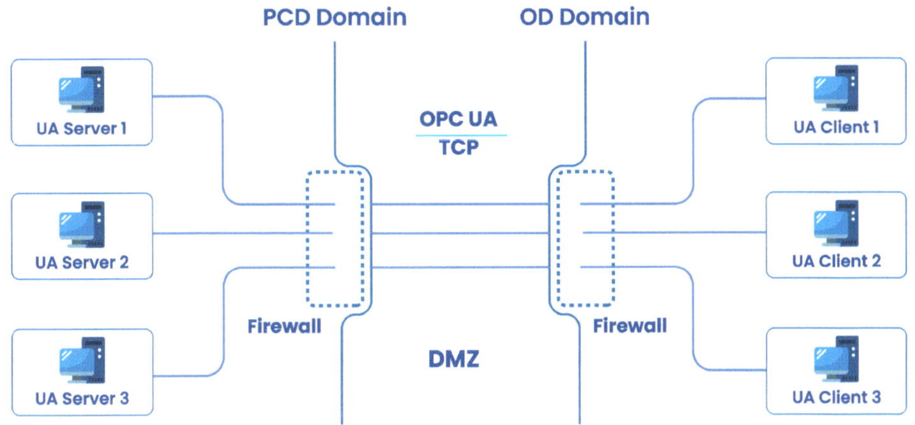

Illustration: Final Architecture with OPC UA Client and Servers Acorss DMZ

CHANGE MANAGEMENT AND WORKFORCE EDUCATION

The transition to a new technology, such as OPC UA, entails not only a transformation of systems but also a transition towards more integrated and data-driven operations. It requires a shift in mindset, culture, and workforce roles.

For instance, with increased data availability, people in decision-making roles will need to recalibrate their approach and learn to incorporate this data in their decisions. Similarly, managers on the factory floor may need to learn how to interpret new data to understand what's happening in real-time to troubleshoot any issues. The IT teams on the other hand will need to understand the new security features and how to set up and maintain the system. The Instrumentation managers must deep dive into the new protocol and its implications for the hardware and software in use with respect to interoperability, platform independence, and integrated security.

In addition to changes in existing roles, new roles may emerge with the implementation of OPC UA. For instance, an OPC UA Administrator role may be required to manage and configure the OPC UA servers and ensure secure and efficient data exchange. Data Analysts might find their roles expanding, as OPC UA can make more data accessible for analysis, potentially leading to new insights and efficiencies.

A move to OPC UA is not just about implementing new software but about adopting new ways of thinking, working, and interacting with data and a workforce that is ready and equipped to use that technology effectively. To handle a shift of this magnitude requires effective change management and workforce education. Some key aspects to consider in this regard are:

- **Clear Communication and Awareness Campaigns to Minimize Resistance:** Help the people involved understand the need behind the shift to OPC UA, the benefits it will bring, and how it will affect individual roles and the company. Encourage questions and feedback from all parties to ensure everyone understands the value of the transition.

- **Get Leadership Buy-in** – Do focused discussions and meetings with leaders at all levels, including managers and supervisors to create appreciation for the outcomes OPC UA can generate for them. Show them the proof of concept to help them appreciate the ROI of the investment. Get them on board and train them to address their team's concerns and support the transition.

- **Invest in Workforce-education and Skill-building to Create Proficiency**: New ways of working and new systems will create new roles in the organization. Train your people to fulfill these new requirements not just with formal training sessions but also with ongoing support, resources, and opportunities for practice.

Create periodic training and refresher programs for new employees to develop proficiency in managing these systems.

Ultimately, the full potential of OPC UA will be realized only when all stakeholders work in tandem, and the organization seamlessly transitions to the new way of working.

OPC UA Pub/Sub
Light- weight
Messaging
OPC UA TCP-IP
Reliable & Secure

OPC UA Myth Busting #8

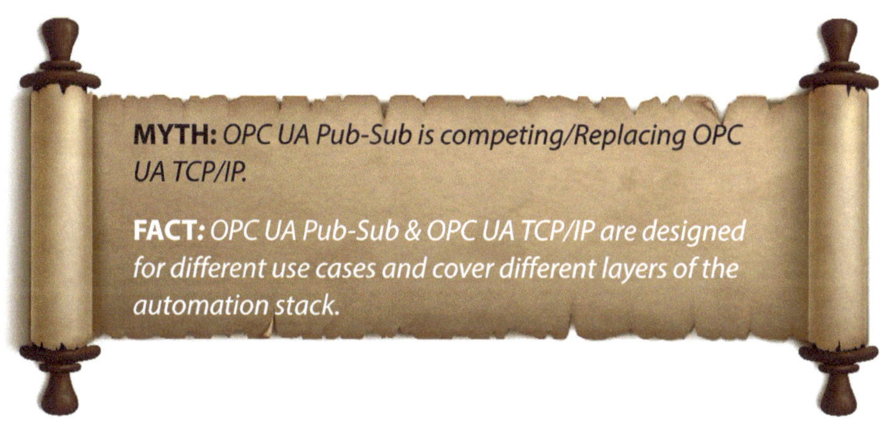

The myth that OPC UA Pub-Sub is competing with or replacing OPC UA over TCP/IP stems from a misunderstanding of the respective roles and capabilities of these two communication methodologies. The aggressive marketing of OPC UA Pub-Sub since its introduction may have contributed to this misunderstanding giving the impression that OPC UA Pub-Sub is being positioned as a successor to OPC UA TCP/IP. Another factor contributing to this myth is the inherent confusion that arises from the fact that both OPC UA Pub-Sub and OPC UA TCP/IP are part of the broader OPC UA framework. Given that they both fall under the OPC UA umbrella, it's not surprising that some people may have mistakenly assumed that they serve the same purposes or that one is intended to be an upgrade of the other.

However, OPC UA TCP/IP and OPC UA Pub-Sub are designed to handle very different use cases. OPC UA TCP/IP is generally used for client-server communication, where two parties maintain a continuous communication line. On the other hand, OPC UA Pub-Sub is designed for scenarios where data needs to be disseminated quickly to multiple recipients, enabling a one-to-many or many-to-many communication model. OPC UA TCP/IP and OPC UA Pub-Sub target also different layers of the automation stack.

Rather than being in competition, OPC UA TCP/IP and OPC UA Pub-Sub should be seen as complementary technologies. Each has unique strengths that make it suitable for specific scenarios.

Interoperability
Data Exchange
Smart Systems
Cyber Security
Automation
Beyond Automation

Moving Beyond Industrial Automation

We've seen how OPC UA has revolutionized industrial automation, streamlining communication between various machines, devices, and software. However, the potential of OPC UA extends far beyond the factory floor. Its cross-platform operability, robust security protocols, and flexible data modelling are finding applications in various sectors.

One of the reasons OPC UA is seen as a protocol only suitable for industrial automation lies in its origins. When the Oil & Gas players first started adopting OPC DA, the COM/DCOM dependence, vendor-lock-in, etc., made it a very costly proposition. But today, with OPC UA's information modeling capabilities, the framework has become very lucrative. And now other industries are waking up to its potential. For instance, managing massive buildings with all their varied smart systems such as HVAC, lighting, security etc. and multiple devices is a tough task. But OPC UA with BACnet makes building automation easier. For instance, a building's HVAC system can adjust temperatures based on data from occupancy sensors and weather forecasts. Similarly, lighting systems can interact with security systems to deter potential intruders. These interactions can result in energy savings, improved security, and a better overall user experience.

In addition, the capability of OPC UA to handle unstructured data – such as PDF, images, etc. - opens new use cases such as security surveillance, traffic management, etc. With OPC UA and companion standards, it's possible to bring together a variety of data formats from multiple sources to a common control tower. It was unprecedented and beyond imagination before OPC UA. Now? It's possible.

Canada for example has been using OPC UA for its traffic control systems. The complex traffic network requires real-time data exchange to ensure smooth operation and prevent congestion. OPC UA can enable smart signals to exchange data - traffic volumes, incident reports, and signal timing changes - with each other and a central traffic control system in real time. The resulting system becomes capable of responding dynamically to conditions on the road, reducing congestion and improving overall traffic flow.

Similarly, OPC UA has promising applications in dairy and agriculture. These operations involve various systems – milking/farming equipment, temperature control, feed/irrigation management, weather monitoring, and animal/crop health monitoring. OPC UA can connect these various systems and devices to give a comprehensive picture of their operations. For instance, an automated milking system could share data with a feed management system, ensuring that cows are fed appropriately for their milk production. Similarly, a temperature control system can adjust the greenhouse climate based on data from plant health monitors. Or an irrigation system can regulate the quantity of water based on weather conditions.

Applications of OPC UA are also cropping up in fields that use robotics. For instance, in healthcare, OPC UA can enable robots to understand patient monitoring systems, or in logistics, warehouse robots can interact with inventory management software.

In conclusion, while OPC UA has made a name for itself in industrial automation, its potential applications are boundless. Its abilities to provide secure, efficient data exchange across a variety of platforms make it an attractive solution in a world increasingly reliant on smart, interconnected systems. As we move towards a more connected future, OPC UA will play a pivotal role far beyond its current contribution in industrial automation in various sectors.

Role of OPC UA in OPAF Standard

Open Process Automation™ Standard (O-PAS™ Standard) or "Standard of Standards" as it's popularly known is an initiative to create a new age automation system with a different architecture than the existing process automation systems that uses Distributed Control Systems (DCS) and Programmable Logic Controllers (PLCs). As automation applications require ultra-high availability and real-time performance, process automation systems have always been highly proprietary.

Open Process Automation™ Standard encompasses multiple individual systems:

- Manufacturing execution system (MES)
- Distributed control system (DCS)
- Safety instrumented systems (SIS)
- Input/output (I/O) points, programmable logic controllers (PLCs), and human-machine interface (HMIs)

In 2016, The Open Group launched the **Open Process Automation™ Forum** (OPAF) to create an open, secure, and interoperable process control architecture to:

- Facilitate access to leading-edge capacity
- Safeguard asset owner's application software
- Easy integration of high-grade components

- Use an adaptive intrinsic security model

- Facilitate innovation value creation

This section of the book highlights why and how OPC UA can be applied to realize the Open Process Automation™ Standard. Before that, let us be familiar with the Open Process Automation™ Forum. In simple terms, The Open Group Open Process Automation™ Forum is an international forum that comprises users, system integrators, suppliers, academia, and organizations.

These stakeholders work together to develop a standards-based, open, secure, and interoperable process control architecture called Open Process Automation™ Standard or O-PAS™. In version 1 of O-PAS™, published in 2019, the critical quality attribute of interoperability was addressed. In version 2, published in January 2020, the O-PAS™ Standard addressed configuration portability, and version 3.0 will be addressing application portability.

Version 1.0 of the O-PAS™ Standard unlocks the potential of emerging data communications technology. Version 1.0 was created with significant information from three existing standards:

- ANSI/ISA 62443 for security

- OPC UA from IEC as IEC 62541 for connectivity

- DMTF Redfish for systems management

The seven parts that makeup the latest preliminary 2.1 version of O-PAS™ Standard are:

- Part 1 - Technical Architecture Overview

- Part 2 - Security (informative)

- Part 3 - Profiles

- Part 4 - Connectivity Framework (OCF)

- Part 5 - System Management

- Part 6 - Information Models based on OPC UA (Multipart specification ranging from 6.1 to 6.6)

- Part 7 - Physical Platform

Part 1 - Technical Architecture Overview

This informative part demonstrates an OPAS-conformant system through a set of interfaces to the components.

Part 2 - Security

This part addresses the cybersecurity functionality of components that should be conformant to O-PAS™. This part of the standard also explains the security principles and guidelines incorporated into the interfaces.

Part 3 - Profiles

This part of the version defines the hardware and software interfaces for which OPAF needs to develop conformance tests and ensure the interoperability of the products. A profile describes the set of discrete functionalities or technologies available for each DCN. They may be composed of other profiles, facets, as well as individual conformance requirements.

Part 4 – O-PAS™ Connectivity Framework (OCF)

This part forms the interoperable core of the system, and OCF is more than a network. OCF is the underlying structure that enables disparate elements to interoperate as a system. This is based on the OPC UA connectivity framework.

Part 5 - System Management

This part covers the basic functionality and interface standards that allow the management of system functions using a standard interface. The system management addresses the hardware, operating systems, and platform software, applications, and networks.

Part 6 - Information and Exchange Models

This part defines the common services and the common information exchange structure that enable the portability of applications such as function blocks, alarm applications, IEC 61131-3 programs, and IEC 61499-1 applications among others.

Part 7 - Physical Platform

This part defines the Distributed Control Platform (DCP) and the associated I/O subsystem required to support O-PAS™ conformant components. It defines the physical equipment used to embody control and I/O functionality.

O-PAS™ Standard version 2.0:

The O-PAS™ Standard supports communication interactions within a service-oriented architecture. In automation systems, it outlines the specific interfaces of the hardware and software components used to architect, build, and start-up automation systems for end-users.

Why OPC UA is important for Open Process Automation™ Forum?

The lower L1, L2 layers of the automation pyramid is heavily proprietary with a tight vendor control over the devices where the PLC's, DCS, sensors, actuators and IO devices operate. This is where the vendors have strong hold over the end-users. As a revenue generating path, they are reluctant to lose this advantage. Additionally, this poses interoperability, security and connectivity issues causing significant lifecycle and capital costs for the stakeholders.

This inherent lack of standardization in the lower OT layers is a constant pressure point for the automation industry. O-PAS™ Standard solves this standardization & connectivity issue and uses OPC UA as one of the foundations for developing this standard. This de-facto standard is used for open process automation integrating controls, data, enterprise systems and serves as a fundamental enabler for manufacturers.

Building the basic components of this standard (like DCN, gateways, OCI interfaces, OCF) using OPC UA helps them achieve secure data integration and interoperability at all levels of the IT/OT integration. This involves leveraging the OPC UA connectivity (Part 4 of O-PAS™ and information modeling capabilities (Part 6 of O-PAS™) which play a key role in the O-PAS™ reference architecture.

How O-PAS™ leverages OPC UA

From the below architecture diagram, it's evident that a Distributed Control Node (DCN) is the heart of the OPAF architecture. Here a single DCN is similar to a small machine capable of controlling, running applications, and other performing functions for seamless data exchange with the higher Advanced Computing Platform (ACP) layers. This component interfaces with the O-PAS™ Connectivity Framework (OCF) layer that is based on the OPC UA connectivity framework.

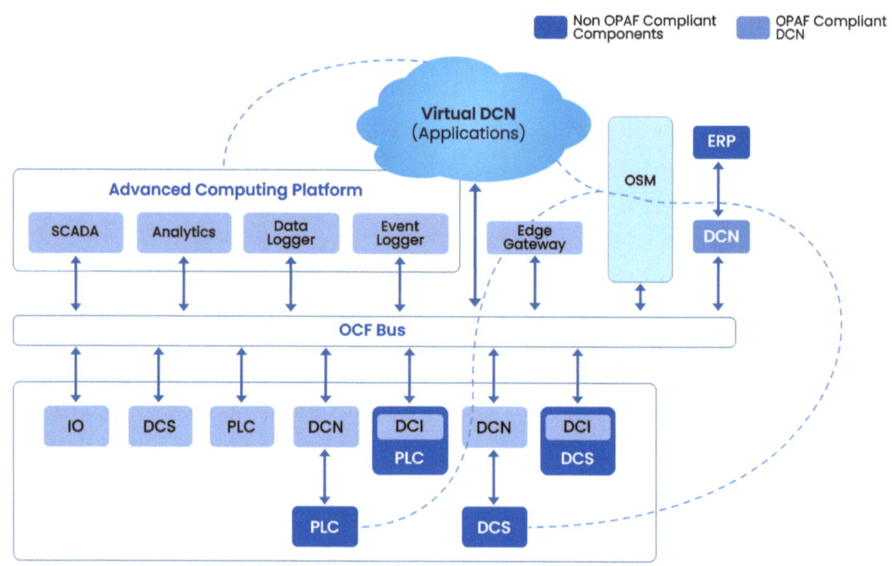

Illustration: High Level OPAF Architecture

The connectivity framework allows interoperability for process-related data between instances of DCNs. It also defines the mechanisms for handling the information flow between the DCN instances. The framework defines the run-time environments used to communicate data.

Basically, each DCN has a profile which describes a set of full-featured definition of functionalities or technologies. For example:

- DCN 01 Profiles (Type - IO + Compute)

- DCN 04 Profiles (Type – Protocol Convert + DCN Gateway)

The DCNs (i.e. O-PAS conformant components) are built conforming to anyone of the primary profiles specified in the O-PAS™:

The OPC UA information model capability is used to define and build these DCN profiles. Part 6 of the O-PAS™ and its subparts defines related set of information and exchange models, such as basic configuration, alarm models, or function block models. This provides a standard format used for the exchange of import/export information across management applications. It also provides standard services used for the download/upload of information to O-PAS™ conformant components.

According to the report **OPC UA Momentum Continues to Build** published by the ARC Advisory Group and endorsed by the OPC Foundation, it provides timely insights into what makes OPC UA the global standard of choice for industrial data communications in process and discrete manufacturing industries. From an IIoT and Industry 4.0 perspective, the report examines how the OPC UA technology is the standard that solves the interoperability challenges.

Key take-away from the report that help maximize OPC UA adoption include:

- OPC UA standard is open and vendor agnostic, and the standard and Companion Specifications are freely available to everyone

- OPC UA is an enabler for next-generation automation standards that will, potentially change the industry structure of process automation e.g. Ethernet Advanced Physical Layer (Ethernet APL), NAMUR Open Architecture, and the Open Process Automation Forum (OPAF)

- OPC UA is arguably the most extensive ecosystem for secured industrial interoperability

- OPC UA is independent of underlying transport layers. As such, it uses the most suitable transports for the right applications (ex. TCP, UDP, MQTT, and 5G)

- OPC UA is highly extensible via its Information Modeling (IM) capabilities. This makes OPC UA an excellent fit for use by automation vendors and other standards organizations wishing to express and share semantic data seamlessly across all verticals

- The OPC Foundation Field Level Communications (FLC) Initiative is defining a new OPC UA Field eXchange (OPC UA FX) standard that is supported by virtually all leading process automation suppliers

- OPC UA FX will extend OPC UA to the field level to enable open, unified, and standards-based communications between sensors, actuators, controllers, and the cloud

- Forward-looking companies should make OPC UA a crucial part of their long-term strategies today because the changes this technology brings will become a necessity faster than most people anticipate

[*Source:* *https://www.automation.com/en-us/articles/june-2021/opc-ua-most-important-interoperability-technology*]

OPAF is making outstanding records in creating a comprehensive, open process automation standard. Since it is partially built on other established industry standards like OPC UA, the O-PAS™ Standard can improve interoperability in industrial automation systems and components.

Glossary

OT	Operational Technology
DMZ	De-militarized Zone
OPC classic	An Old technology of data exchange in OPC which worked over COM-DCOM
OPC D.A	The current Values Server with in OPC Classic
OPC UA	The latest OPC Technology for Data Exchange which is platform and technology independent
OPC UA Pub-Sub	A flavor of OPC UA which works on Publish-Subscribe method instead of Request Response
COM-DCOM	Propitiatory Microsoft Technology for efficient Data Exchange
OPAF	Open Process Automation™ Forum
AMQP	Advanced Message Queuing Protocol
MQTT	Message Queuing Telemetry Transport
TCP/IP	Transmission Control Protocol/Internet Protocol
ROI	Return on Investment

About the Authors

Praveen Kumar Singh
Utthunga

Praveen Kumar Singh has over two decades of industrial experience, with extensive exposure to OPC technologies. With a unique focus on the end-use community in implementing OPC solutions, Praveen has also earned the globally recognized authority of OPC UA-Accredited Expert. A certified OPC trainer, cybersecurity professional, and a system architect, Praveen is also credited with three patents in OPC technologies. His customer-centric approach strongly complements his technical expertise, enabling him to craft diverse OPC solutions for various industrial applications. At Utthunga, Praveen heads the OPC UA vertical, delivering world-class solutions and advisory across a wide spectrum of industrial domains. In his leisure time, Praveen enjoys exploring the natural beauty of uncharted terrains and writing poems!

Smitha Rao
Utthunga

Smitha Rao is a OPC Foundation accredited expert with 20 years of experience in Industrial Automation domain, especially the Process Automation industry.

She is responsible for strategic association with consortiums like OPC Foundation, TheOpenGroup, FieldComm Group, FDT Group, ODVA and PNO. To her credit, she is an editor of few of OPC UA companion specifications. She has also been part of research in supporting EtherNet/IP in FDI specification and authored IEEE publications in OPC, FDT/FDI & EtherNet/IP technologies. Smitha is cofounder of Utthunga, MD of Utthunga GmbH and has been Awarded the "Women Entrepreneur of the Year" by Govt of India. Beyond the professional realm, she is an ardent adventurist with a penchant for skydiving and scuba diving, coupled with a love for biking, arts, and crafts.

Chatrapathi G V
Utthunga

Chatrapathi G V, an OPC Foundation-accredited expert, brings over two decades of expertise in Industrial Communication Protocols and IIoT solutions. As an ODVA Designated Trainer and a recognized expert in PROFIBUS & PROFINET, he's made significant contributions to various OPC UA specifications and prototyping initiatives. With a rich history of working on OPC UA PubSub over TSN and FX, Chatrapathi has designed and developed groundbreaking technical solutions to foster strategic business insights for many leading Industrial OEMs worldwide.

Andreas Faath
VDMA

Andreas Faath is the Managing Director of VDMA Machine Information Interoperability and a member of the OPC Foundation's Board of Directors. He plays a pivotal role in championing interoperability activities driven by industry specifications and technologies for the Mechanical Engineering industry. He represents the global mechanical and plant engineering sectors in Interoperable Information Exchange and Digital Transformation. With a Mechanical and Plant Engineering background, Andreas worked as a scientist at the Technical University of Darmstadt, focusing on Digitalization and IIOT. Later, he delved into R&D for process optimization and digitalization in vehicle testing at General Motors.

References

[1] https://www.history.com/this-day-in-history/fords-assembly-line-starts-rolling

[2] https://www.mckinsey.com/industries/industrials-and-electronics/our-insights/unlocking-the-industrial-potential-of-robotics-and-automation

[3] https://www.globenewswire.com/en/news-release/2023/03/14/2626969/0/en/Industrial-Automation-Market-Predicted-to-Garner-USD-493-billion-by-2032.html

[4] https://www.mckinsey.com/capabilities/operations/our-insights/the-imperatives-for-automation-success?ref=marketsplash.com

[5] https://marketsplash.com/automation-statistics/#link6

[6] https://www.grandviewresearch.com/industry-analysis/industrial-automation-market?ref=marketsplash.com

[7] https://www.accenture.com/pl-en/insights/technology/interoperability

[8] https://www.mckinsey.com/capabilities/operations/our-insights/the-imperatives-for-automation-success?ref=marketsplash.com

[1] https://www.automationmedia.com/Port1050/OPCDownloads/Extract%20from%20Book%20OPC%20Unified%20Architecture.pdf

[i] https://opcfoundation.org/news/press-releases/arc-advisory-group-report-finds-opc-ua-standard-well-positioned-base-iiot-i4-0-solutions/

[ii] https://opcfoundation.org/wp-content/uploads/2023/05/OPC-UA-Interoperability-For-Industrie4-and-IoT-EN.pdf

Reference Links for Illustrations

Illustration #	Reference Link
1	https://opcconnect.opcfoundation.org/2016/03/reshape-the-automation-pyramid/
2	Industry 4.0 development timeline and enabling technologies. | Download Scientific Diagram (researchgate.net)
3	https://theautomization.com/opc-ua-vs-da/
4	https://theautomization.com/opc-ua-vs-da/
5	https://opcfoundation.org/about/opc-foundation/history/
6	https://youtu.be/Bz0xkq97ox8?si=NhwS-WiK9Bu5AEro
7	https://support.industry.siemens.com/cs/document/109773628/what-is-the-difference-between-variable-type-and-data-type-in-the-type-system-of-opc-ua-?dti=0&lc=en-AO
8	https://bit.ly/47hexgO
9	https://www.theseus.fi/bitstream/handle/10024/800870/Sehli_Taha.pdf?sequence=2
10	https://www.theseus.fi/bitstream/handle/10024/800870/Sehli_Taha.pdf?sequence=2
11	https://reference.opcfoundation.org/Core/Part1/v104/docs/6.3.7
12	https://www.automaatioseura.fi/site/assets/files/1550/f2068.pdf
13	https://documentation.unified-automation.com/uasdkhp/1.4.1/html/_l2_ua_discovery_connect.html
14	https://documentation.unified-automation.com/uasdknet/2.6.1/html/L2UaDiscoveryConnect.html

15	https://opcfoundation.org/wp-content/uploads/2023/05/OPC-UA-Interoperability-For-Industrie4-and-IoT-EN.pdf
16	https://opcfoundation.org/wp-content/uploads/2023/05/OPC-UA-Interoperability-For-Industrie4-and-IoT-EN.pdf
17	https://www.slideshare.net/sadatullazishan/opc-ua-inside-out-part-6-brownfield-and-greenfield-webinar
18	https://www.slideshare.net/sadatullazishan/opc-ua-inside-out-part-6-brownfield-and-greenfield-webinar
19	https://bit.ly/3SsYWXt
20	https://umati.org/wp-content/uploads/062343_umati_info-brochure_web.pdf
21	https://reference.opcfoundation.org/Machinery/v103/docs/8.1

Notes

www.ingramcontent.com/pod-product-compliance
Lightning Source LLC
Chambersburg PA
CBHW040920110726

48006CB00001B/3